Taunton's
BUILD LIKE A PRO™
Expert Advice from Start to Finish

BUILDING a SHED

BUILDING a SHED

JOSEPH TRUINI

The Taunton Press

Text © 2002 by Joseph Truini

Photographs © 2002 by Geoffrey Gross, except where noted

Illustrations © 2002 by The Taunton Press, Inc.

The Taunton Press
Inspiration for hands-on living®

The Taunton Press, Inc., 63 South Main Street, P.O. Box 5506, Newtown, CT 06470-5506

e-mail: tp@taunton.com

EDITOR: Andrew Wormer

COVER AND INTERIOR DESIGN: Lori Wendin

LAYOUT: Jeff Potter/Potter Publishing Studio

ILLUSTRATOR: Mario Ferro

Taunton's Build Like a Pro® is a trademark of The Taunton Press, Inc.,
registered in the U.S. Patent and Trademark Office.

Library of Congress Cataloging-in-Publication Data

Truini, Joseph.

　Building a shed : expert advice from start to finish / Joseph Truini.

　　p. cm. -- (Taunton's build like a pro)

Includes index.

　ISBN-13: 978-1-56158-619-6

　ISBN 1-56158-619-6

　1. Sheds--Design and construction--Amateurs' manuals.　I. Title. II.

Series.

　TH4962 .T78 2002

　690',89--DC21

　　　　　　　　　　　2002151662

Printed in the United States of America

15　14　13　12　11　10

The following manufacturers/names appearing in *Building a Shed* are trademarks: Cedar Shake and Shingle Bureau®,
CERTI-SPLIT™, Chicago Metallic®, Dek-Block™, DEKBRANDS™, Dura Slate Roofing System™, Fast Framer™,
Georgia Pacific®, James Hardie Building Products®, Lee Valley Tools®, Natural Select™, NatureWood®, Preserve®,
Shingle Shield™, SharkSaw®, Simpson Strong-Tie®.

Construction is inherently dangerous. Using hand or power tools improperly or ignoring safety practices can lead to
permanent injury or even death. Don't try to perform operations you learn about here (or elsewhere), unless you're certain
they are safe for you. If something about an operation doesn't feel right, don't do it. Look for another way. We want you to
enjoy the craft, so please keep safety foremost in your mind whenever you're in the shop.

To my father, Joseph Truini, Sr., who taught me that working with your hands—whether building a house or writing a book—is a noble cause.

To Marla, Kate, and Chris, for keeping me afloat.

Acknowledgments

Writing a book is a lot like building a shed: It always takes longer than you think, but in the end, it's well worth the effort. It's also virtually impossible to do without the help of others.

Thank you to all the hard-working, talented people at The Taunton Press. I'm proud to have my name on one of your books. Thanks especially to executive editor Helen Albert, who first offered me this book project and showed confidence in my ability from the very beginning. I'm also grateful to associate managing editor Jennifer Renjilian for her professional guidance, Wendi Mijal for her skilled art direction, and Jenny Peters for her always friendly, helpful assistance.

Thanks, too, to book editor Andrew Wormer, for his keen eye, helpful suggestions, and professionalism throughout the arduous task of molding my words.

I was extremely fortunate to have worked with photographer Geoffrey Gross, whose technical skill, immeasurable patience, and dedication to this project can't be overstated. The words may be mine, but it's Geoffrey's images that truly make this book special.

I could never have produced this book without the help and unflagging support from Pete and Arlene Charest and Jack Quintiliani, owners of Better Barns, a Connecticut-based shed-building company. They were always open to suggestions and showed an inordinate amount of patience throughout every design and construction phase of this project. Thanks, too, to their talented carpentry crew, Greg Butkus, Paul Clough, and Dave Scarles, for never running out of patience or good humor, and to Brautigam Land Surveyors, P.C. of Newtown, Connecticut, for their help in surveying building sites.

Finally, a debt of gratitude is owed to John Blaney and Bill Jenks, building officials in Litchfield County, Connecticut, who expertly and patiently answered all my arcane (and sometimes inane) questions about building codes, zoning regulations, and frost heave. Their contributions made this a much better book.

Contents

Introduction

AS THE SON OF a carpenter, I had no choice but to learn my father's craft. It was passed down to me through the same complex genetic code that gave me his dark eyes and stubborn nature. My apprenticeship began at the tender, clueless age of 12, when my father woke me early one summer morning (much to my surprise) and announced that it was time to go to work. As I climbed into his truck, sleepy-eyed and silent, I had no idea that I was embarking on a journey of discovery and purpose that would guide me through life to this very day.

During that summer, and subsequent summers for the next several years, I worked alongside my dad and learned a lot about carpentry, electrical wiring, plumbing, and life. Later, as a young adult, I used those skills to earn a living as a remodeling contractor, union carpenter, cabinetmaker, and, eventually, writer.

For the past 20 years, I've written about homebuilding, remodeling, and woodworking for nearly every possible medium: books, newspapers, websites, television, radio, and several magazines, including *Popular Mechanics, Today's Homeowner, Country Living, Handy, Home Mechanix,* and *This Old House.* It was during my years as a magazine editor that I first began writing about backyard outbuildings and discovered that nearly every homeowner— regardless of the size of his or her house— needed a storage shed. Over the years, I've written articles about and built more than a dozen sheds. Therefore, when the fine folks at The Taunton Press asked me to write this book, I was confident that my experiences in the building and publishing fields would prove invaluable. I hope you think so, too.

From the very beginning, the goal of this book has been to provide you with the information, inspiration, and confidence to build your own backyard storage shed. And not just one of the structures shown in this book, but any shed at all. Like most other shed books, this one has a lot of photos of attractive outbuildings and many construction drawings, but that's where the similarities end. This book gives specific information on what you need to know before you build a shed, including design considerations, building code issues, and evaluating your storage needs. A chapter devoted to shed-building techniques covers everything from foundations to roof framing. Another chapter explains the wide variety of shed-building materials, including siding, roofing, doors, and windows.

However, what makes this book truly unique is that each of the final four chapters shows how to build a particular shed from scratch. These structures were specifically designed and constructed for this book and include an easy-to-build 2-ft. by 6-ft. Shed Locker that's perfect for storing lawn and garden tools; a charming 8-ft. by 12-ft. Saltbox Potting Shed, which features a cedar-shingle roof and traditional Dutch door; a 10-ft. by 16-ft. Colonial-Style Garden Shed that combines classical design with beautiful vertical-board cedar siding; and a spacious 12-ft. by 20-ft. Gambrel Storage Barn, which has double-wide sliding doors and an interior storage loft.

For each shed, there are dozens of step-by-step photos and precise drawings to guide you through each phase of the construction process. However, if you need additional information, mail-order plans are available for each of the four sheds.

Finally, it took about a year to produce the book you now hold in your hands, but it—and all that I am today—really started early one summer morning more than 35 years ago.

How to Use This Book

IF YOU'RE READING THIS, you're a doer who is not afraid to take on a challenging project. We've designed this book and this series to help you do that project smoothly and cost effectively.

Many doers jump in and do, reading the directions only if something goes wrong. It's much smarter (and cheaper) to start by knowing what to do and planning the process step by step. This book is here to help you. Read it. Familiarize yourself with the process you're about to undertake. You'll be glad you did.

Planning Is the Key to Success

This book contains information on designing your project, choosing the best options for the results you want to achieve, and planning the timing and execution. We know you're anxious to get started on your project. Take the time now to read and think about what you're about to do. You'll refine your ideas and choose the best materials.

There's advice here on where to look for inspiration and how to make plans. Don't be afraid to attempt drawing your own plans. There's no better way to get exactly what you want than by designing it yourself. If you need the assistance of an architect or engineer, you'll find advice in this book on why and how to work with those professionals.

After you've decided what you're going to undertake, make lists of materials and a budget for yourself, both of money and of time. There's nothing more annoying than a project that goes on forever.

Finding the Information You Need

We've designed this book to make it easy to find what you need to know. The main part of the book details the essential parts of each process. If it's fairly straightforward, it's simply described. If there are key steps, they are addressed one by one, usually accompanied by drawings or photos to help you see what you will be doing. We've also added some other elements to help you understand the process better, find quicker or smarter ways to accomplish the task, or do it differently to suit your project.

Alternatives and a closer look

The sidebars and features included with the main text are there to explain aspects in more depth and to clarify why you're doing something. In some cases, they are used to describe a completely different way to handle the same situation. We explain when you may want to use that method or choose that option, as well as detail its advantages. The sidebars are usually accompanied by photos or drawings to help you see what the author is describing. The sidebars are meant to help, but they're not essential to understanding or doing the process.

Heads up!

We urge you to read the "Safety First" and "According to Code" sidebars we've included. "Safety First" gives you a warning about hazards that can harm or even kill you. Always work safely. Use appropriate safety aids and know what you're doing before you start working. Don't take unnecessary chances, and if a procedure makes you uncomfortable, try to find another way to do it. "According to Code" can save you from having trouble with your building inspector, building an unsafe structure, or having to rip your project apart and build it again to suit local codes.

There's a pro at your elbow

The author of this book, and every author in this series, has had years of experience doing this kind of project. We've put the benefits of their knowledge into quick tips that always appear in the left margin. "Pro Tips" are ideas or insights that will save you time or money. "In Detail" is a short explanation of an aspect that may be of interest to you. While not essential to doing the job, it is meant to explain the "why."

Every project has its surprises. Since the author has encountered many of them already, he can give you a little preview of what they may be and how to address them. And experience has also taught the author some tricks that you can only learn from being a pro. Some of these are tips, some are tools or accessories you can make yourself, and some are materials or tools you may not have thought to use.

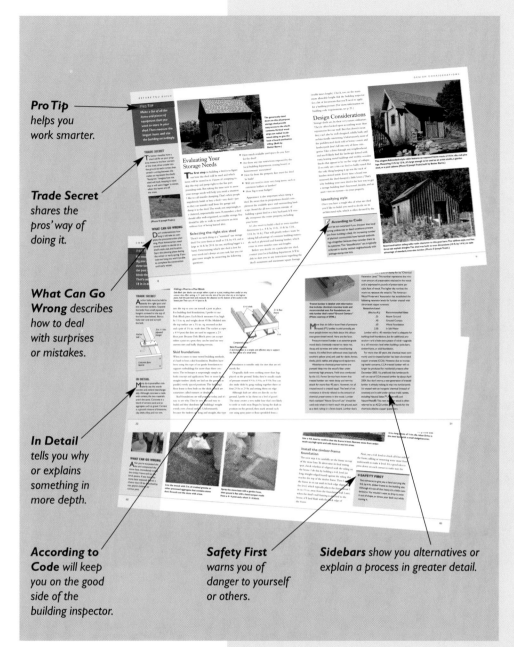

Pro Tip helps you work smarter.

Trade Secret shares the pros' way of doing it.

What Can Go Wrong describes how to deal with surprises or mistakes.

In Detail tells you why or explains something in more depth.

According to Code will keep you on the good side of the building inspector.

Safety First warns you of danger to yourself or others.

Sidebars show you alternatives or explain a process in greater detail.

Building Like a Pro

To make a living, a pro needs to work smart, quickly, and economically. That's the strategy presented in this book. We've provided options to help you make the best choices in design, materials, and methods. That way, you can adjust your project to suit your skill level and budget. Good choices and good planning are the keys to success. And remember that all the knowledge and every skill you acquire on this project will make the next project easier.

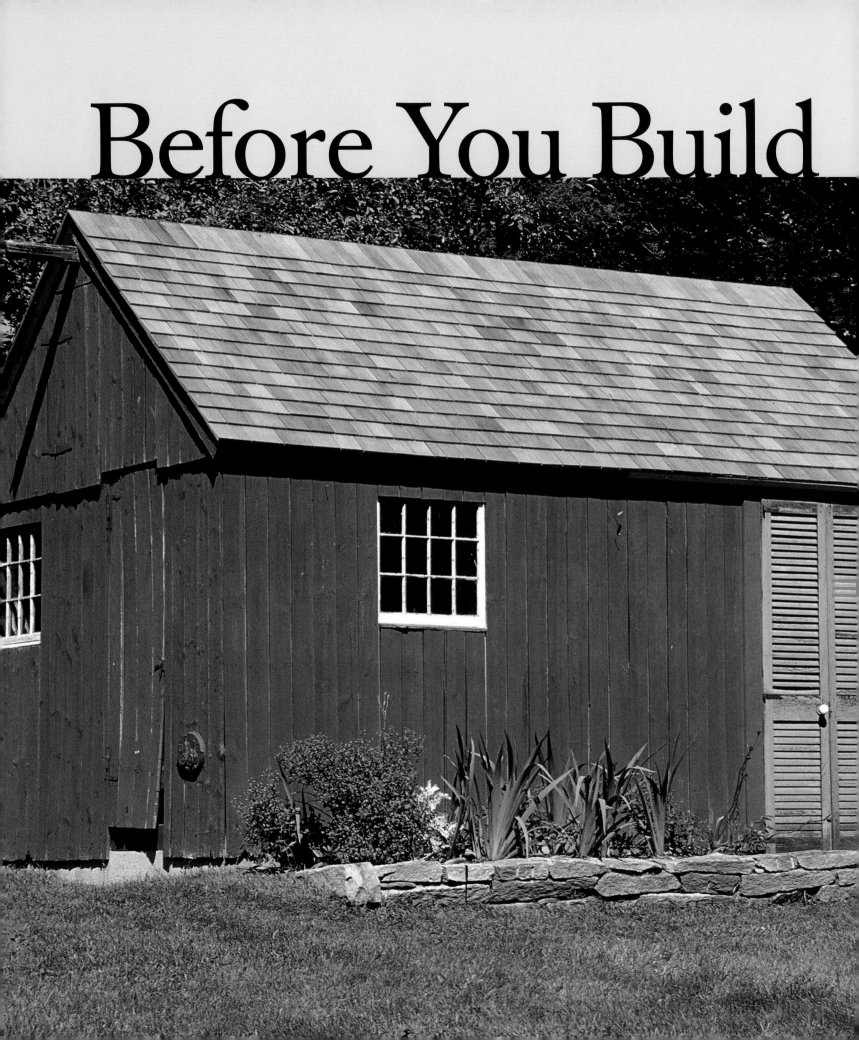

Before You Build

CHAPTER ONE

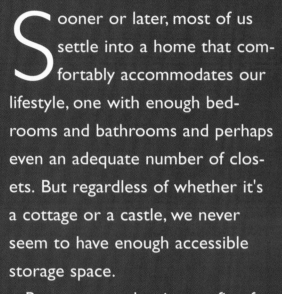

Sooner or later, most of us settle into a home that comfortably accommodates our lifestyle, one with enough bedrooms and bathrooms and perhaps even an adequate number of closets. But regardless of whether it's a cottage or a castle, we never seem to have enough accessible storage space.

Basements and attics are fine for keeping holiday decorations and old clothing, but a separate outbuilding is the only practical solution for storing garden tools, lawn mowers, camping gear, bicycles, ladders, lumber, and all the other stuff we typically cram into the garage or pile up outdoors.

The challenge, however, isn't simply to build a shed, it's to build the right shed. With a little forethought and planning, it's easy to determine which size and style shed is best for you and your family. And you can begin by answering this simple question: What are you going to use the shed for?

TRADE SECRET

To better visualize how a shed will fit on your property, measure the four corners of the shed, drive a stake into the ground at each corner, then stretch a string between the stakes to represent the shed's "footprint." Imagine how the shed will look, keeping in mind that it will seem bigger in winter, when the leaves are off the trees.

(Photo © Joseph Truini.)

WHAT CAN GO WRONG

Don't underestimate how long it will take to complete the preconstruction planning. Most homeowners need several weeks to decide on a shed's style, size, and location. Start this planning phase during the winter or early spring. If you wait too long, you won't be able to complete the construction until early winter.

The generously sized doors on this all-purpose storage shed provide easy access to the shed's contents. Vertical wood strips are nailed to plywood siding to give the look of board-and-batten siding. (Built by Better Barns.)

Evaluating Your Storage Needs

The first step in building a shed is to figure out how the shed will be used and which items will be stored in it. I know it's tempting to skip this step and jump right to the fun part: pounding nails. But taking the time now to assess your storage needs will help you avoid a situation I like to call transfer dumping. That's when people impulsively build or buy a shed—any shed—just so they can transfer stuff from the garage and dump it in the shed. The result, not surprisingly, is a cluttered, impenetrable mess. Remember, a shed should offer well-organized, accessible storage. You should be able to walk in and retrieve an item without fear of being buried alive.

Selecting the right size shed

There's no such thing as a "standard" size storage shed. I've seen them as small as 2 ft. by 4 ft. and as large as 16 ft. by 20 ft. (to me, anything bigger is a barn). Determining which size shed is best for your needs isn't always an easy task, but you can gain some insight by answering the following questions:

- How much available yard space do you have for the shed?
- Are there any size restrictions imposed by the local building department, zoning board, or homeowners' association?
- How far from the property line must the shed be built?
- Will you need to store very long items, such as extension ladders or lumber?
- How big is your budget?

Appearance is also important when sizing a shed. Be aware that its proportions should complement the available space and surrounding landscape. Avoid the all-too-common mistake of building a giant shed in a tiny backyard. It'll visually overpower the entire property, including your home.

It's also smart to build a shed to even-number dimensions (i.e., 8 ft. by 10 ft., 10 ft. by 12 ft., 12 ft. by 16 ft.). That will greatly reduce waste by taking full advantage of common building materials, such as plywood and framing lumber, which come in even-number sizes and lengths.

Before you decide on a particular size shed, contact your local building department. It'll be able to alert you to any restrictions regarding the shed's minimum and maximum square footage (width times length). Check, too, on the maximum allowable height. Ask the building inspector for a list of documents that you'll need to apply for a building permit. (For more information on building code requirements, see p. 21.)

Design Considerations

Storage sheds are, by their very nature, utilitarian. They're often looked upon as nothing more than repositories for our stuff. But that doesn't mean they can't also be well-designed, solidly built, and architecturally interesting. Unfortunately, most of the prefabricated sheds sold at home centers and lumberyards don't fall into any of those categories. Take a drive through any neighborhood and you'll likely find the landscape dotted with rusty, leaning metal buildings and wobbly wooden shacks that appear to be on the verge of collapse. (I recently saw a tin-can shed so badly rusted that the only thing keeping it up was the stack of lumber stored inside. Every time a board was removed, the shed slumped a little lower.) That's why building your own shed is the best way to get a storage building that's functional, durable, and an asset—not an eyesore—to your property.

Identifying style

Once you have a rough idea of what size shed you'd like to build, you need to decide on its architectural style, which is often dictated by the

---

✓ According to Code

Don't be too surprised if you discover that local zoning ordinances or deed covenants prevent you from building a shed. An increasing number of planned communities have banned outbuildings altogether because they consider them to be eyesores. This "beautification" act is typically enforced in thickly settled neighborhoods with postage-stamp-size lots.

This elegant Adirondack-style cabin features an exposed frame made of 10-in.-dia. red-pine logs. Measuring 12 ft. by 12 ft., it's large enough to be used as an artist studio, a garden shed, or a pool cabana. (Photo © Joseph Truini; built by Better Barns.)

Board-and-batten siding adds rustic character to this goat barn. The saltbox-style roof has three-tab asphalt shingles. The shed was built to even dimensions (10 ft. by 14 ft.) to take advantage of standard, even-size lumber. (Photo © Joseph Truini.)

PRO TIP

Neither aluminum nor vinyl siding is rugged enough to absorb the inevitable hard knocks and rough handling that every outbuilding endures.

TRADE SECRET

Can't decide on which type of siding to use for your shed? Call a local lumberyard and get price quotes for each type. Costs vary widely, and price alone may be the deciding factor. Remember: You can't build a redwood shed on a plywood budget.

Availability also plays a key role, which is why it's important that you place your siding order early. Lumber, like produce, isn't always available every day of the year. Your construction schedule can easily be delayed by two or more weeks if the siding you desire is suddenly unavailable.

WHAT CAN GO WRONG

Windows are a great addition to nearly any shed. They add architectural interest and help brighten up the interior storage space. But many first-time shed builders make the mistake of installing too many windows. The result is a building that looks too spotty, with far too much glass and not nearly enough visible siding. From a practical aspect, too many windows means very little wall space for hanging tools, installing cabinets, and mounting shelves.

roof design and siding material. Most sheds have a simple gable roof, though saltbox, gambrel, and shed roofs (like the roof on the Lean-to Shed Locker in Chapter 4) are other options.

When it comes to siding the exterior walls of a shed, you also have several options. Wood is by far the most commonly used material, which isn't surprising, as it's readily available, affordable, and easy to work with. Choices in wood include bevel siding (a.k.a. clapboards), shingles, vertical boards, and plywood. Plywood siding isn't quite as stylish as other types of wood siding, but it's less expensive and goes up very quickly, which is why it's often used on larger sheds and barns.

Fiber-cement siding—a relatively new shed-building material—offers a practical alternative to wood clapboards. It looks like wood but is made of fiber-reinforced cement, so it's extremely durable and completely fireproof.

Window wisdom

Windows also influence a shed's style, but to a slightly lesser degree than the roof or siding does. When designing your shed, remember that the size and placement of the windows should be in proportion to the building. That may sound obvious, but I've seen dozens of poorly designed sheds—mostly cheap, prefab models—that have one tiny window on one wall. The puny port-hole allows in very little light and virtually no air, and it only cheapens the look of the building.

You don't have to be Frank Lloyd Wright to figure out which size windows will look best; all you need to do is open your eyes. Look around your neighborhood or local lumberyard to find a shed that's about the same size as the one you're planning to build. Stand back far enough to view the entire building and surrounding area. Look for a pleasing balance between the amount of glass and the visible siding. The windows should be properly proportioned, neither too big nor too small for the structure. When you find a well-proportioned shed, measure the size and placement of its windows.

Here's another equally effective method that doesn't require you to even leave the house: Get a ruler and make a scaled drawing of the building. Then sketch in the windows, adjusting their dimensions until they look well balanced. For example, if two small, square windows don't look right, then enlarge them. If that doesn't help,

This elegant English potting shed has a small porch, vertical-board siding, and faux-slate roof shingles. Western-red-cedar siding was chosen for its good looks and long life. (Photo © Joseph Truini; built by Better Barns.)

The 9-ft.-tall sidewalls of this spacious storage barn allowed the builder to install a series of 2-ft. by 3-ft. windows above the doorway, which flood the interior space with natural light. (Photo © Joseph Truini.)

replace the two windows with a single large, rectangular one.

Maintaining proper proportions applies to all parts of the building, including the exterior trim around the windows and doors. Generally speaking, smaller windows and doors look best with narrow trim—that is, trim less than 3 in. wide. Large windows and wide doorways can accommodate much wider trim boards.

Keep in mind, too, that you're not required to put windows in every shed. Many small outbuildings—say, 8 ft. by 8 ft. or smaller—often look fine without any windows.

Interior features

As you're working your way through the initial design and planning stage, think about customizing the interior of the shed, too. Even if you're going to use the shed for only general storage, chances are good that you'll need at least one or two shelves and a few hanging hooks.

A wooden workbench is a popular shed accessory, but build one only if you're sure you'll be able to keep the floor clear of clutter. It's very easy for a shed to become so crammed full of stuff that you can't even see the workbench.

If you plan to store bulk goods in the shed, such as fertilizer, potting soil, or pet food, be sure

Optional Shed Upgrades

Regardless of the size or style of shed you plan to build, consider adding one or more of the following upgrades to your building:

- A ramp, especially if you'll be storing a riding mower, wheelbarrow, or bicycle.
- Extra windows for a brighter interior. (Keep in mind, however, that for each window you install, you'll sacrifice a little wall-hung storage space.)
- A storage loft above the ceiling joists. This is an excellent, out-of-the-way place to store lumber and large boxes.
- Extra-wide doorways for driving in large lawn tractors, riding mowers, garden tillers, and snow throwers.
- Adjustable, wall-mounted shelves for neatly organizing tools and supplies.
- Built-in workbenches; they're especially useful for potting plants and woodworking.
- Pegboard for neatly hanging tools on the walls.
- Gable or ridge vents, which allow hot, stale air to escape.
- Gutters to divert rain away from the doorway.
- Window screens to keep out insects, especially bees and mosquitoes.
- A cupola with a weather vane for a bit of country elegance, especially on larger sheds.
- Window shutters and flower boxes for a cozy cottage look.

Pegboard panel provides a quick, easy way to keep tools readily accessible and neatly organized.

PRO TIP

Can't decide which color to paint or stain your shed? Color photocopies of the building plan with felt-tip markers to create various color schemes.

TRADE SECRET

Looking to add a workbench to your shed but don't quite have the space for one? Try this space-saving solution: Make a 16-in.- to 24-in.-wide by 6-ft.- to 8-ft.-long benchtop out of ¾-in. plywood. Measure 36 in. up from the floor and attach the benchtop to the wall with fold-down metal brackets (available at any hardware store for about $15 apiece). The collapsible brackets allow you to lower the benchtop flat against the wall when it's not in use.

(Photo © Joseph Truini.)

Here's one way to neatly organize a shed: Place a workbench and a shelving unit in opposite corners, use hanging hooks to utilize all available wall space, and, most important, keep the floor in the middle of the shed clear of clutter. (Photo courtesy Handy Home Products.)

to allow space for a few large plastic storage bins. The covered bins are also useful for organizing aluminum cans, glass bottles, and other recyclables.

Selecting a style

Choosing a particular shed style is often based on personal preference and taste, but there are also some practical concerns. For example, sheds with gable roofs offer tall sidewalls that are useful for putting up shelves and hanging long-handled tools. The tall walls can also accommodate larger windows and doors. However, the steeply sloping roof doesn't provide much space above the ceiling joists or a lot of headroom near the walls.

Gambrel-style roofs—sometimes referred to as barn roofs—are quite spacious inside and offer plenty of headroom. But their shorter sidewalls offer very little space for hanging tools or mounting shelves. It's also difficult to install doors on the sides of most gambrel-style sheds because of the limited wall height. Some builders solve these problems by building taller sidewalls. That way, they gain usable wall space without sacrificing headroom. In fact, many barns are built this way, but they are large enough to accommodate the extra wall height; most sheds are not. Simply

This 7-ft. by 9-ft. plywood garden shed has a side hutch for storing recycling bins. Free building plans are available from Georgia Pacific®.

building taller walls on a shed will throw the entire building out of proportion and make it look oddly stretched. In spite of their limited wall space, gambrel-roof sheds are popular, simply because they look like a traditional barn.

If you're not sure which style of shed you want to build, there are several places to get ideas. Look through home-improvement magazines and cut out photos of outbuildings that you like. Get brochures and catalogs from shed manufacturers. Visit a local shed builder or home center and tour the model buildings on display (don't forget to take along a camera and tape measure). Check out newsstands for magazines that sell homebuilding plans; many of them also show photos or illustrations of sheds, garages, and other outbuildings.

A Working Plan

Once you've decided on the size and style of the shed you want to build, you need a building plan. There are two basic ways to obtain one: buy an existing plan or draw one from scratch.

Plan sources

Mail-order shed plans are available from books, magazines, websites, tool catalogs, and companies that sell homebuilding plans (see Reources on p. 198). Such plans typically cost between $10 and $50, depending on the size and complexity of the building.

Buying a ready-made plan will save you the trouble of designing and drawing a shed from scratch, which is a rather time-consuming process. However, drawing your own plan allows you to create a unique shed that's specifically designed to suit your tastes and storage needs; it also costs less. A third option is to buy a mail-order plan, then alter the design to satisfy your requirements.

Looking for a better deal? Many building-product manufacturers, such as Georgia Pacific and Simpson Strong-Tie®, offer free plans in the

This handsome 10-ft. by 14-ft. shed features a post-and-beam frame made with 6×6 timbers; it was built from plans by Mercurial Editorial. (Photo © Smith-Baer.)

hope that you'll then buy their lumber or hardware.

Drawing a design

If you can't buy a plan of the exact shed you'd like to build, then you'll have to draw it yourself. Now don't panic. You don't have to be a skilled draftsperson to create an accurate building plan. All you need is some ¼-in.-scale graph paper and a few basic drawing instruments, such as a mechanical pencil, a ruler, dividers, a protractor, a compass, and an eraser.

Start by drawing an elevation view of each exterior wall of the shed. This is simply a flat, one-dimensional view of the wall, as it would look if you were standing directly in front of it. Be sure to include the roof and the location of the doors and windows. Draw the shed using a ½-in. scale. In other words, each ¼-in. square printed on the

WHAT CAN GO WRONG

The cost of building a typical shed foundation is minimal—if you have a level lot free of boulders and trees. However, the cost can skyrocket if the site needs extensive excavation. Ask the building department which type of foundation is required for your outbuilding. It's good to know early on whether your budget must be expanded to cover the expense of digging footing holes and pouring concrete piers.

TRADE SECRET

Prevent resinous knots from bleeding through the top coat of paint by spot-priming all the knots. Apply a stain-killing primer with a brush or roller—depending on the number of knots—and allow it to dry completely before applying two top coats of paint. Spot-priming can be used for the siding and all exterior trim, such as fascia boards, friezes, soffits, and corner boards.

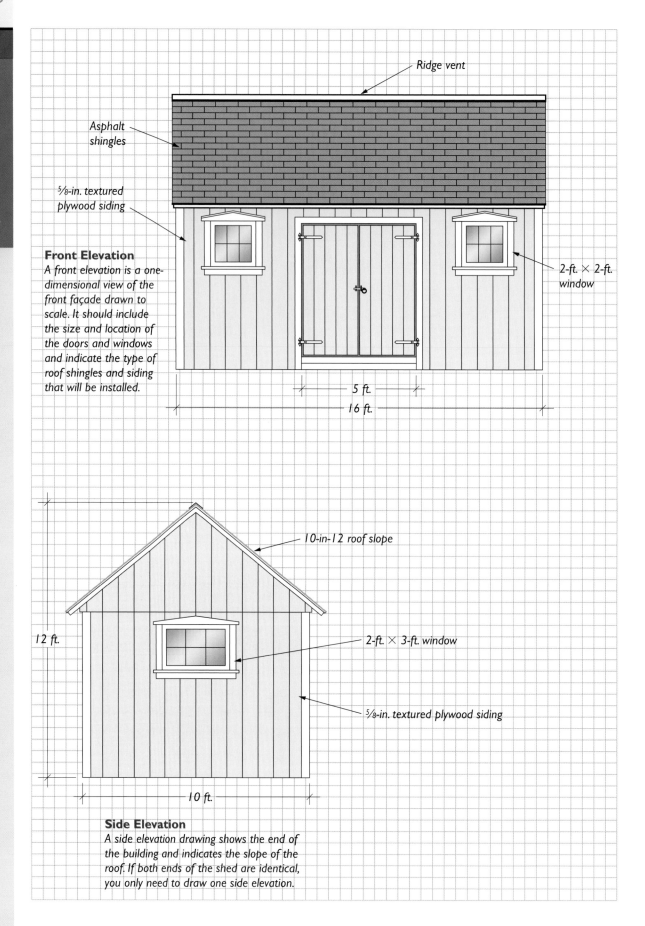

Ridge vent

Asphalt shingles

⅝-in. textured plywood siding

Front Elevation
A front elevation is a one-dimensional view of the front façade drawn to scale. It should include the size and location of the doors and windows and indicate the type of roof shingles and siding that will be installed.

2-ft. × 2-ft. window

5 ft.

16 ft.

10-in-12 roof slope

12 ft.

2-ft. × 3-ft. window

⅝-in. textured plywood siding

10 ft.

Side Elevation
A side elevation drawing shows the end of the building and indicates the slope of the roof. If both ends of the shed are identical, you only need to draw one side elevation.

paper is equal to 6 in. That means 1 ft. of actual distance is represented by two squares, or ½ in. By the way, you can buy ½-in.-scale graph paper, but the smaller squares in the ¼-in. grid allow you to draw 6-in. dimensions much more accurately.

The reason it's important to draw the elevations to a precise scale is that it's the easiest way to tell whether the shed will be properly proportioned. If any part of the building is even slightly out of whack, it'll be clearly visible in the drawings. Here are some things to look for:

- Does there appear to be too much space between the windows? Enlarge them, add shutters, or install another window in between.

- Is the shallow roof pitch making the shed look a little squashed? Redraw the roof at a steeper angle.

- Does the shed look too long and narrow? Decrease its overall length or make the building wider.

- Don't be afraid to experiment with different designs. Making adjustments to the shed at this point is easy—just pull out your eraser and try again.

After completing the elevation drawings, sketch a plan view (a.k.a. bird's-eye view) of the proposed foundation. Be sure to include the exact spacing and alignment of the foundation system, which can be concrete blocks, footing holes, wooden skids, or poles. For specific information on foundations and a wide variety of shed-building techniques and materials, see Chapters 2 and 3.

Sweat the details

You typically don't need to draw a separate framing plan of how the floor, walls, and roof will be framed, but include this information somewhere

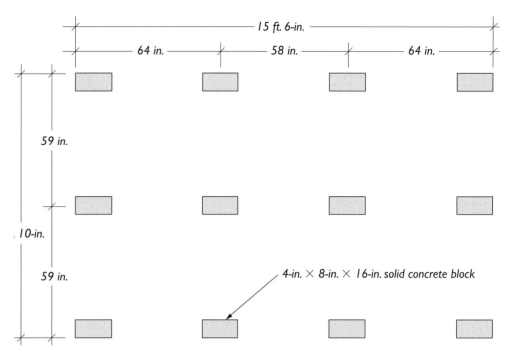

A Typical Foundation Plan
A foundation plan is a bird's-eye view of the proposed foundation. This drawing is of an on-grade foundation made up of 12 solid-concrete blocks. A foundation plan must include the size of the blocks and the spacing between them.

A Common Design Dilemma

Should a shed have the same siding, paint color, and roof shingles as the house? One commonly applied rule states that if an outbuilding is close to the house—say, 25 ft. or less—it must be the same material and color as the house. The rationale? If a nearby shed is architecturally different from the house, it'll somehow look odd or out of place. I don't know who came up with this rule or when it became a shed-building commandment, but it's time to admit that there's no logic behind it.

A well-designed shed can enhance any property, regardless of its paint color, siding material, or proximity to the house. In fact, the most attractive outbuildings are typically ones with their own distinctive look. Now, with that said, you should know that there's no reason why you can't build a shed to resemble your house. It's just that there's no design rule that states you absolutely must.

PRO TIP

When building a large shed, consider a door at each end or a pair of doors in the middle of a sidewall. You'll reach items stored all the way in the back more easily.

TRADE SECRET

If you're building your shed on a slightly sloping plot of land, position the front of the shed (the side with the primary door) on the high side of the slope. That way, the land will slope down toward the rear of the shed and the slanting grade won't be so obvious. Placing the door on the high side will also make it much easier to step into the shed because the door's threshold will be closer to the ground.

WHAT CAN GO WRONG

Is it possible to build on top of a low, wet spot? Yes, if you absolutely must, but it probably won't be easy or inexpensive. In most cases, you'll have to dig up the entire site and install a drainage system of gravel, landscape fabric, perforated pipes, and extra topsoil.

The shed-roof extension on this white-painted outbuilding provides valuable storage space for lawn mowers and bikes. The painted finish lends a more formal look to the building. (Photo © Crandall & Crandall.)

on the drawing. Be sure to list nominal dimensions and names of the parts, along with the layout spacing and sheathing. For example, for the walls you might add the following construction detail: 2×4 wall studs spaced 16 in. on center and sheathed in ½-in. exterior-grade plywood. It's necessary to include all the major construction details because most building departments require you to submit this information along with the plan when you apply for a building permit. Again, see Chapter 2 for various building techniques for each construction phase.

Finishing School

The exterior finish you choose for a shed also influences its style. There are three basic choices: paint, stain, and clear wood preservative. Personal preference plays a key role in which finish you choose, but your selection will also depend on the look you want, your siding material, and the amount of maintenance you want to undertake.

Why paint?

Paint comes in the widest variety of colors and lends a more formal look to an outbuilding. However, keep in mind that painted wood will eventually need to be scraped and repainted, and most people would rather vacuum Texas than scrape paint.

If you're going to paint a shed, be sure to use a high-quality, 100% acrylic latex paint. In most cases, you'll also have to apply a primer, unless the siding or trim boards come with a factory-applied coat of primer. Check the label on the can of top coat paint for specific priming requirements. It's also a good idea to paint the interior walls, which will help stop moisture from passing through the walls from the inside and blistering the paint off the outside. Consider, too, the siding material. If you're using a beautiful wood, such as cedar, redwood, or cypress, do you really want to hide the attractive wood grain behind a coat of paint?

Stain options

Semitransparent stain is an excellent choice for maintaining wood's natural texture. It won't peel or blister the way paint will, but stain will fade over time, especially in areas exposed to direct sunlight. In addition, semitransparent stains are typically available in only wood-tone colors and various shades of gray.

Solid-color—or opaque—stains are formulated to fall somewhere in between paint and semi-transparent stain. They come in a wider variety of colors than semitransparent stains do, but because they contain more "solids," they can peel and blister the way paint does.

A popular shed-finishing technique is to use semitransparent stain on the body of the shed and solid-color stain—in a contrasting color—on the trim.

Clear finishes

Clear wood preservatives require the least amount of maintenance because they don't peel or fade. Although a preservative won't prevent the wood from slowly weathering to a silvery gray color, it will help protect the shed from the harmful effects of sunlight, rain, snow, and mildew. When buying a wood preservative, look for one that contains water repellent and mildewcide.

If you can't decide on a finish, you should at least apply a coat of clear wood preservative to the shed. Exterior wood should never be left unprotected. If you apply a preservative, you'll still be able to paint the shed some time in the future. Just be sure to scrub the siding clean before applying a coat of primer and a top coat of paint.

Siting the Structure

The next step in the preliminary planning stage is to select an appropriate building site for the shed. This may sound like a simple task—and if you're lucky, it just might be—but if you pick the wrong spot, you'll end up with an unusable outbuilding.

There are several factors that influence site selection, including the size and topography of the property and how you plan to use the shed. Of course, the local authorities will have their say, too. The building inspector must approve both the building plan and the proposed site before he or she will grant a building permit. In most towns,

Want to add a little style and architectural interest to your shed? Try adding a Dutch door, a window box, and louvered shutters. Paint adds a more formal look than a stained or natural finish. (Photo courtesy Walpole Woodworkers.)

permission to build must also be granted by the inland wetland commission, health department, and zoning board. Here are several important issues to consider as you walk your property in search of the perfect shed site.

Spots to avoid

It's never smart to build a shed at the bottom of a hill or in a low-lying area where water collects. The excessive moisture can rot wood, blister paint, and cause hinges to rust. It'll also promote mold and mildew growth on items stored in the shed. Plus, the ground near the shed will turn into a soggy, muddy quagmire after every rainstorm.

Some of the most attractive sheds I've seen are tucked deep into the woods, completely surrounded by towering trees and lush ground cover. Unfortunately, these lovely woodland settings are ill-suited for storage buildings. First of all, the trees and surrounding vegetation block out much of

PRO TIP

Before going to the building inspector for a permit, first get approval from the inland wetlands commission, health department, and zoning board.

How High Is the Roof?

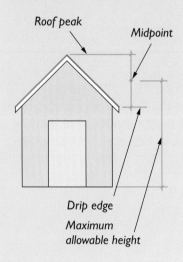

Roof peak

Midpoint

Drip edge

Maximum allowable height

TRADE SECRET

There's a little-known building code that concerns the height of an outbuilding, but it also affects the foundation. Any outbuilding that exceeds the maximum allowable height, which is typically 12 ft., must have a permanent foundation. Most people would assume that the code refers to the shed's height as measured from the ground to the roof peak. But to many building departments, the maximum height is actually the distance from the ground to the midpoint between the roof peak and the drip edge along the bottom of the roof.

This 8-ft. by 10-ft. storage shed was converted into a children's playhouse. It features an arched Dutch door and a giant 30-lite window. Avoid siting a shed in the middle of a wooded lot, where it will prematurely age. (Photo courtesy Walpole Woodworkers.)

the sun and wind, so the shed receives very little sunshine or ventilation. As a result, the building remains dark and damp, creating the perfect environment for mold and mildew growth.

The shed's roof is especially susceptible and can quickly become overgrown by a thick layer of moss, lichen, mildew, or vines. It's not unusual to find grass, mushrooms, or even small trees growing out of the roof. If you have to mow the roof, you picked the wrong spot for your shed.

Woodland sheds are also under constant assault from falling branches, acorns, leaves, pine needles, and other types of canopy debris. Furry forest creatures are much more likely to move into or under a shed built in the woods. And there's always the potential for severe damage caused by a fallen tree. It takes only one direct hit from a large tree to reduce a shed to a pile of kindling.

Wooded sites require a lot of extra prep work, too. You'll have to cut down trees, pull up ground cover, and scrape the forest floor clean and level. In some cases, you'll even have to dig out boul-

Measuring just 6 ft. by 8 ft., this compact saltbox has board-and-batten siding, three-tab asphalt roof shingles, and a weather vane cupola. Siting it near trees—not in the woods—takes advantage of a woodland setting but protects the shed from falling branches and mildew growth. (Photo courtesy Walpole Woodworkers.)

ders and stumps, which is not a fun way to spend a weekend.

To avoid these problems but still enjoy a lovely woodland setting, build the shed at the edge of the forest. You can even set the building back into the woods a couple of feet, but don't bury it completely.

Setback distances

Setback requirements are another important consideration when siting a shed. These are strict town ordinances that establish how far away the shed must be from such things as:

- Side, front, and rear property lines
- Streets, driveways, and sidewalks
- Houses, garages, and decks
- Swimming pools and ponds
- Septic tanks and leach fields
- Wetland areas
- Water-well pumps

- Power lines and telephone poles
- Easements

Setback distances vary widely from town to town, even from one neighborhood to the next. They can range anywhere from 10 ft. (leach field) to 100 ft. or more (wetlands). Again, check with the local zoning board or building department for specific information, and be sure to adhere to the letter of the law. If you violate one of the setbacks, the town can legally make you move the shed, which, as you can imagine, is no easy chore.

What if the only decent building site in your entire yard is 10 ft. away from the side property line and the setback requirement is 15 ft.? Are you out of luck? Not necessarily. Most towns let you apply for a variance, which means you have a chance to appear in person before the zoning commission and plead your "hardship." Its members will hear your case and decide whether or

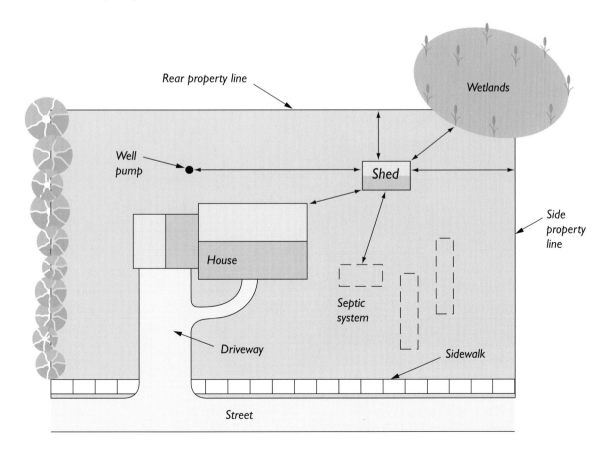

Setback Requirements
Setback distances indicate how far away the shed must be from such things as wetlands, septic tanks, leach fields, property lines, well pumps, and other structures. Check with the local zoning board for the setback requirements in your town. They can range anywhere from 10 ft. or 15 ft. up to 100 ft. or more.

PRO TIP

Don't allow the lumber delivery truck to drive across your lawn during or immediately after a rainstorm. It'll tear up the grass and severely compact the wet soil.

WHAT CAN GO WRONG

How are you going to get your building materials to the building site? That isn't a problem if the site is next to your driveway, but what if it's located deep in the backyard, through the fence, past the swimming pool, and beyond the garden? If the lumber delivery truck can't reach the site from your driveway, try a neighbor's driveway or a side street. If that isn't feasible, call your buddies and grab a wheelbarrow; you'll have to haul the lumber piece by piece.

TRADE SECRET

Before renting a piece of equipment, such as a gas-powered trencher or a stump grinder, ask around the neighborhood to see whether anyone else needs to use it. Then you can split the cost among more people and save a considerable amount of money.

Looking more like a country cottage than a toolshed, this stucco building has an inviting front porch and two gabled dormers. (Photo © Crandall & Crandall.)

The Phoenix Solar Shed is part storage shed, part greenhouse. The skylights are placed on a south-facing slope to capture the most amount of sunlight. This shed is available as an easy-to-build kit from Handy Home Products. (Photo © Jerry Kolesar.)

not to grant the variance. It can be denied for any number of reasons, including a protest from a neighbor, health and safety concerns, environmental issues, and the ever-changing wind of local politics.

If the variance is granted, you'll be able to build the shed closer than the rule normally allows but—and this is important—no closer than the variance states. In my hypothetical example, the shed could now be 10 ft. from the property line, but not any closer.

The mere thought of having to apply for a variance causes great angst in many homeowners. Don't let this happen to you. Zoning boards are made up of fellow homeowners who are typically very supportive and sympathetic to variance requests. In fact, all the building officials I interviewed said they grant a vast majority of the variances brought before them. One building inspector said that consent is given more than 90 percent of the time.

However, zoning boards often meet only once a month; if the docket is filled for next month's meeting, you'll have to wait until the following month to plead your case. Avoid long construction delays by finding out early on from a zoning board member whether you'll need to apply for a variance.

You'll also most likely have to pay a variance application fee. How much? That depends on the town. I know a homeowner who had to shell out $100 for a variance to build a shed, but that wasn't the worst of it: The zoning commissioner also required her to hire a licensed surveyor to draw the shed's location on the plot plan of the property. That cost another $800—ouch!

Easy access

Sheds are often relegated to the deepest corner of the yard, far from the house. And there's nothing wrong with that, if its primary use is to store a riding mower or engage in a specific activity, such as potting plants or woodworking.

However, if it will be an active storage building that your family will use on a regular basis, situate it close to the house or garage. You'll then be able to retrieve an item from the shed without having to traipse across the backyard, a great convenience if you live in an area that gets a lot of snow or rain. And the kids will be much more likely to put away their bikes, toys, and sports equipment if the shed is close by.

Sunny outlook

To gain the most amount of natural light, position the shed so the wall with the largest windows or widest doorway faces south. Now, if you're using the shed for general storage, its orientation to the

sun isn't all that important. But for potting sheds and other specific uses, the sun's solar energy and natural light can help grow plants and illuminate workbenches. To help control the amount of light and solar heat entering the shed, outfit the windows with roller shades or operable wooden shutters.

Code Concerns

It's common knowledge that before you can build anything on your property, including a shed, you must contact the local building department and apply for a building permit. However, few homeowners realize that the building department should be their last stop at city hall when requesting permission to build. That's because the building inspector won't grant a permit until you get approval from the inland wetlands commission, health department, and zoning board. Here's a detailed look at the typical chain of events.

Inland wetlands

Call your town clerk or city hall and ask whether there's an inland wetlands commission. If there is, you must submit a plot plan of your property showing the proposed building site of the shed. If you don't have the plot plan, check the files at the building department. Not there? Then you'll have to hire a surveyor to draw one.

The minimum setback distance from a wetland is usually 100 ft. You may not have wetlands on your property, but you'll need to have that fact verified by the commission. Technically, wetlands are specific types of soil, as designated by the Department of Environmental Protection. Wetlands include obvious areas, such as swamps, lakes, ponds, rivers, and streams. They also encompass places that carry intermittent or seasonal water from any number sources, including rain runoff from a nearby property or street.

How do you know whether you have wetlands on your property? It should be noted on the ori-

+ SAFETY FIRST

It's seldom necessary to install electricity in a shed that's strictly used for storage, but for safety's sake, consider lighting the way to the shed. Place a floodlight on the house or garage and aim it at the shed. It will prevent you from tripping over or walking into something during nighttime trips to the building.

TRADE SECRET

Excavating contractors don't always like to transport heavy equipment for small jobs, such as clearing a shed site. That's why many of them have a four-hour minimum, which is a lousy deal if you have only an hour or two of work. One way around this problem is to hire a local landscaper, arborist, or farmer who does land clearing as a side business. Such freelancers typically have a no-job-too-small policy and very reasonable rates, especially if they live close by.

WHAT CAN GO WRONG

When a professional contractor fails to show up on the day promised, it can throw off your construction schedule and spoil an entire weekend that you had planned to work on the shed. Avoid this problem by hiring the contractor as early as possible. Call at least a month ahead of time, and get him or her to commit to a date. Then call again one week before the scheduled date to confirm the time of arrival. Remember, it's the squeaky wheel that gets the contractor.

A trellis screen made of crisscrossing 1×2s supports a lush crop of creeping vines. Note that the trellis is mounted 4 in. from the siding. Be aware that you're not allowed to build on property that has been designated as wetlands. (Photo © Crandall & Crandall.)

ginal plot plan of the lot, but only if your home was built after the formation of the inland wetlands commission. If your home is more than 15 or 20 years old, it probably predates the commission and the plot plan won't indicate any wetland areas. However, that doesn't mean they don't exist. If necessary, hire a state-licensed soil scientist to check your property for wetlands.

Health department

If your home has a septic system, the health department will want to make sure you don't build the shed too close to the septic tank or leach fields. But more important, it wants to make sure you maintain something known as a "viable reserve area."

Septic systems sometimes fail, and when they do, they must be replaced. The town will want to make sure there's enough open land on your property to install a new, often bigger septic system should your current one fail. This reserve area must also be large enough to permit heavy equipment to come in and install the new tank

and leach fields. Therefore, the shed can't be built near the existing septic system or within the viable reserve area.

If you don't have a septic system but have city water and sewer instead, you have a little more flexibility regarding where to build the shed. However, you still must get the municipal water company to sign off on your plot plan, so be sure to call the water company and ask about any building-site restrictions.

Zoning board

The inland wetlands commission and health department tell you where you can't build. Go to the zoning board to find out where you can put the shed.

As mentioned earlier, local zoning boards establish setback distances and handle variances. As long as the shed site doesn't violate any of the setbacks, there's no reason for the board to deny your request to build. If necessary, you can always apply for a variance.

Building department

After you've run the gauntlet and received approval from the three offices listed previously, bring your shed plans to the building department. The building inspector's job is to make sure the shed conforms to all local and national building codes. He or she will carefully check the foundation type; the grade of plywood specified; and the size and spacing of the lumber used for the floor joists, wall studs, and roof rafters. Be sure to read Chapter 2 for specific information on various shed-building techniques.

The most critical building code concerns the type of foundation required. There will be more on this subject in the next chapter, but be aware that there are two basic types: on-grade and permanent foundations. The code typically permits small to medium-size outbuildings to be supported by an on-grade foundation made up of concrete blocks or weather-resistant wooden timbers set directly on the ground. (The sheds shown in Chapters 4, 5, and 6 all feature on-grade foundations.)

However, bigger sheds—usually those larger than 200 sq. ft. or taller than 12 ft.—require a permanent or frost-proof foundation (see the Gambrel Storage Barn in Chapter 7). For this type, you must dig down past the point where the ground freezes in winter, known as the frost line, and then pour concrete footings or piers. The frost-line depth varies, depending on your region of the country. In cold climates where problems associated with frost heave are most common, the frost-line depth ranges from 36 in. to 48 in. Ask the building inspector for the exact depth in your area.

Every town charges a building-permit application fee. Sometimes it's a standard amount of $15 to $25, but often the fee is based on the overall cost of the outbuilding. For example, the fee might be $5 per $1,000 of cost. Therefore, the application fee for a $3,000 shed would be $15.

The 6-ft.-wide roof extension on this shed is sheathed in lattice so that dappled sunlight can reach the potting benches below. (Photo © Saxon Holt.)

On-site inspections

When you finally receive the building permit, ask whether any inspections are necessary. Most towns require inspections for permanent foundations and for adding electricity to a shed. The building inspector will want to see the footing holes before you fill them with concrete to verify that they're dug to the correct depth.

✔ According to Code

If you hire a professional contractor to build part of your shed, the town may require the contractor to apply for a permit and submit a valid state license to the building department. If the contractor works in your town all the time, his or her license will probably be on file with the building inspector.

WHAT CAN GO WRONG

Keep in mind that besides not being allowed to build on wetlands, you're also not allowed to disturb the area in any way. That includes cutting down trees, removing plants, or digging drainage ditches.

TRADE SECRET

When you find a photograph of a shed that you're interested in building, you need to know only one critical dimension to figure out the rest of the shed's proportions. Lay a ruler across the known dimension, which can be any part of the shed, such as its length, width, or rafter thickness. Note the measurement on the ruler and use that as your scale. For example, if you know the building is 10 ft. wide and it measures 5 in. in the photo, then 1 in. represents 2 ft. of actual space. Knowing which scale to use allows you to calculate other dimensions easily, including door width, wall height, and window size and spacing.

If you're running an underground electrical cable from the house to the shed, the inspector will want to see the trench before you backfill it to confirm that it's deep enough and that you're using the appropriate cable. Inspections are also required for the rough wiring that feeds the wall receptacles (outlets), light fixtures, and switches inside the shed.

Even if your shed doesn't have a permanent foundation or electricity, chances are good that you'll still need to have a final inspection. The building inspector will need to see the completed shed to verify that its size, location, and construction are in accordance with the approved building permit.

Hiring Help

We all need a little professional help from time to time. As I mentioned earlier, the goal of this book is to provide you with the necessary information and confidence to build your own shed. However, that doesn't mean you can't hire a professional contractor for certain aspects of the project.

There's no shame in hiring a pro to tackle jobs that you don't have the time or inclination to do or that you simply can't do because you don't possess the skills, experience, or special equipment. Here are my top three reasons for calling in a contractor:

The big dig

Hire a pro whenever a site requires extensive excavation. A single backhoe can clear and level the average shed site in less than an hour. Using a shovel and rake may take you two or three weekends.

When it comes to digging footing holes for a permanent foundation, you have a couple of options. If the soil is soft and easy to dig, you can save yourself some cash by using a manual post-hole digger or renting a gas-powered auger. However, if there's even the slightest chance that you'll run into rock or clay, hire an excavator.

Concrete solutions

Mixing and pouring concrete for a few footings or piers isn't all that much work. In fact, it's kind of fun if you've never done it before. In most cases, you won't even need to rent a portable mixer; a wheelbarrow and a hoe will work fine.

However, if the shed has a concrete-slab foundation and you're new to concrete work, you may want to hire a mason. Pouring a slab isn't that difficult, but if it's done incorrectly, it's nearly impossible to fix once the concrete has cured. Plus, it requires a few specialty tools that most homeowners don't own, such as a bull float, a hand edger, and a finishing trowel.

For an additional fee, a mason will also build the wooden form and prepare the site for the pour, but any competent do-it-yourselfer can easily complete those tasks.

Getting wired

Depending on how you plan to use your outbuilding, you may or may not need to wire it with electricity. If it'll be used for general storage, there's really no need to electrify the building. You can use a flashlight for the few times you may need to enter the shed at night. However, if you plan to use the shed as a woodshop, home office, or arts and crafts center, you'll obvious need elec-

Get two sheds in one with this cleverly designed 6-ft. by 16-ft. building from Walpole Woodworkers. On the left is a workshop; to the right, behind the double doors, is general storage. When electrifying an outbuilding, play it safe and hire a licensed electrician. (Photo courtesy Walpole Woodworkers.)

tricity for lighting, power tools, computers, and other electronic equipment.

Ordinarily, power is brought to a shed from the house via an underground cable, which is buried in a trench and hooked up to an existing house circuit. Occasionally, it's easier to run a new circuit from a nearby utility pole. Except for basic tasks, such as installing electrical boxes or hooking up switches and outlet boxes, making electrical connections is work that should be done by a licensed electrician.

You may be able to save the electrician some time—and yourself some money—by doing some of the preliminary work. For example, if the electrician plans to pull power from the house, you can dig the trench and run the cable to the shed. Be sure to dig down to the code-specified depth (usually 18 in.) and use only an underground feed (UF) cable, which is specifically designed for burial. Then have the electrician hook up the cable to the main panel in the house and to the receptacles, switches, and other devices in the shed. Don't forget that the building inspector will want to see the cable and trench before it's backfilled, and inspect all the rough wiring in the shed.

+ SAFETY FIRST

Working with electricity is potentially dangerous, and one mistake can literally kill you. If you're planning to wire your shed with electricity, play it safe: Hire a licensed electrician to bring power to the building and connect the new circuits. You can install the electrical boxes and do the rough wiring, but steer clear of live circuits.

Construction

CHAPTER TWO
Methods

There are many ways to build a storage shed; in this chapter, I'll explore all the possibilities for each construction step, from creating a stable foundation to framing the roof. In many instances, you'll have a choice between three or four building techniques. For example, you can make the shed's floor out of plywood, brick, concrete, or gravel. The walls can be framed with 2×4 studs or 4×4 posts and beams. At the door of the shed, you can build steps, a ramp, or a deck.

The specific techniques that you choose will often be based on personal preference, cost, degree of difficulty, or the topography of the building site. However, be aware that you won't always be the one to choose the construction method. In certain cases, how you build will be determined by the building inspector. For example, if the shed is larger than a certain size, you'll most likely be required to build a frost-proof foundation.

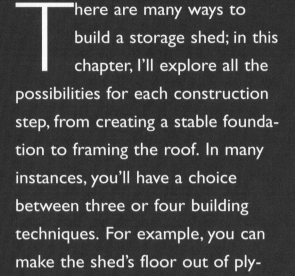

TRADE SECRET

When building a solid-concrete block foundation, it's important that all the blocks be level. However, it's equally important that the blocks in each row be perfectly aligned. The best—and fastest—way to line up the blocks is with a taut string. Install the first and last block in each row. Then stretch a length of mason's line along the edge of the two end blocks and use it as a guide to set the intermediate blocks.

IN DETAIL

Always start an on-grade foundation at the highest spot and work toward the lowest. If the property slopes more than 2 ft. or so over the length or width of the building, you'll have to either bring in some heavy equipment to level the property or choose another building site.

Solid-concrete blocks are ideal for building on-grade foundations. The blocks are aligned in rows and checked for level with a long, straight 2×4 and a 4-ft. level.

Choosing a Foundation

The success or failure of any outbuilding relies heavily on its foundation. No structure—regardless of how well it's designed or built—will survive for very long on a weak or poorly made base. Therefore, choosing and building a proper foundation is the single most important construction step in the entire project.

On-grade vs. frost-proof foundations

Shed foundations fall into two basic categories: on-grade and frost-proof. On-grade foundations (sometimes called "floating foundations") sit right on the ground and are sufficient for all but the very largest outbuildings. They're also the quickest and simplest to build because they don't require you to dig deep holes or pour concrete footings or piers. On-grade foundations are usually made of pressure-treated lumber (see Lean-to Shed Locker on p. 76) or solid-concrete blocks (see Saltbox Potting Shed on p. 106).

Permanent, frost-proof foundations are more difficult to build, but they're by far the strongest and longest lasting. These types of foundations are designed for cold-weather regions where ground movement caused by freeze/thaw cycles can affect a building.

The most popular material for building frost-proof foundations is poured concrete, which can take the form of a footing, pier, or slab. You can also create a permanent foundation—and avoid frost-heave troubles—with an age-old building technique called pole-barn construction. In this system, several tall round poles or square posts set into deep holes support the structure (see Gambrel Storage Barn on p. 168).

The best foundation to build for your shed will largely depend on what the building inspector

Fiber-form tubes are commonly used to build permanent, frost-proof foundations. Each tube is set in a hole dug below the frost line, then filled with concrete.

Pole-barn frames consist of round poles or square posts set in deep holes. Here, a pressure-treated 4×4 post is used.

recommends, but keep in mind that it's often based on three key factors: the shed's size, the region of the country in which you live, and the type of shed floor you desire. To help you choose the best foundation for your shed, let's take a close look at seven foundation systems: four on-grade types and three frost-proof ones.

On-Grade Foundations

It's no surprise that most sheds are designed to be built with an on-grade foundation. This base is quick and easy to build, relatively inexpensive, and adaptable enough to accommodate all but the most severely sloping sites. In addition, the components are small and light enough to easily set into place and shift around, making it very easy to get everything square and level. Although it's not technically a "permanent" foundation, an on-

What Is Frost Heave?

Frost heave sounds like something that happens after you eat too many snow cones. But if you live in a cold-weather region, it's no joking matter. Frost heave is a phenomenon that occurs when moisture in the soil freezes and then expands, pushing the ground—and anything buried in it—upward. When the ground thaws, the soil slumps back down. As you can imagine, all this up and down movement isn't very good for a shed. Frost heave can raise a building 3 in. or 4 in. off the ground, then drop it back down. The problem is, the building seldom settles back into its original, square and level position.

Fine-grained sand and stiff clay soils are more susceptible to frost heave than coarse-grained soils and compacted gravel are, but the phenomenon can occur virtually anywhere, especially if the building site doesn't effectively drain away groundwater.

The best way to prevent this unsettling condition is to dig past the point where the ground freezes—an area known as the frost line—and then pour a concrete pier or footing. How deep you need to dig depends on how far down the frost penetrates the soil in your particular corner of the world. For example, in New England and the Upper Plains, you may need to dig 42 in. or more. In the mid-Atlantic region, 24 in. to 36 in. may be sufficient. Check with the local building department for the exact frost-line depth in your town.

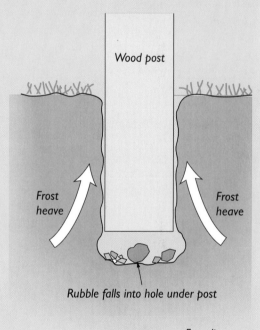

Wood post

Frost heave

Frost heave

Rubble falls into hole under post

Frost line

✓ According to Code

An increasing number of building departments are requiring homeowners to install ground anchors, which prevent a shed from blowing over or sliding off its foundation. The steel-cabled anchors are bolted through the mudsill at each corner of the building. Then a steel rod is used to drive the hold-down spike deep into the ground.

PRO TIP

Gravel and crushed stone are available in 50-lb. bags, but it's cheaper if you buy it by the truckload from a garden shop, nursery, or masonry supplier.

Eight solid-concrete blocks are arranged in two rows to form this on-grade foundation. Identical diagonal measurements indicate a square layout.

WHAT CAN GO WRONG

An on-grade foundation will last a lifetime if the ground below remains dry and undisturbed. Never build on a site that collects standing water. Soggy soil will eventually swallow up the foundation. And don't set the foundation too close to any trees. The roots can grow underneath the shed and lift the foundation right off the ground.

TRADE SECRET

It's often necessary to stack two or more solid-concrete blocks on top of one another to create a level foundation. To keep the blocks from sliding out of position as you set the floor frame in place, use a caulking gun to apply a generous bead of construction adhesive between the blocks. You can also use the adhesive to glue shims to the tops of the blocks.

grade foundation, when properly built, will probably outlast the shed it supports.

Solid-concrete blocks

In this type of foundation, the shed is supported by a series of solid-concrete blocks, which are laid out in straight, evenly spaced rows. The number of blocks needed and the spacing between them is determined by the size of the shed and the lumber used for the floor joists. For example, the 8-ft. by 12-ft. Saltbox Potting Shed (see p. 106) has 2×6 floor joists. Its foundation is made up of eight blocks set in two rows. A 10-ft. by 16-ft. shed requires 12 blocks arranged in three rows.

It's important to note that you must use only solid-concrete blocks for this type of foundation. Standard wall block or any other hollow block will eventually crack and crumble under the weight of the shed. If you have trouble finding solid blocks at a home center or lumberyard, visit a masonry supplier.

The blocks measure 8 in. wide by 16 in. long and come in 4-in.- and 2-in.-thick units. The thicker blocks are placed first, with the thinner "patio" blocks laid on top when you need to raise one block even with the others. In some cases, you may need to stack two or three 4-in. blocks on top of each other to raise the lowest corner

In most cases, you'll need both 2-in.-thick patio blocks and 4-in-thick solid-concrete blocks to build an on-grade foundation.

A shallow bed of gravel placed underneath concrete foundation blocks aids drainage and helps prevent them from sinking into the soil. (Photo © Joseph Truini.)

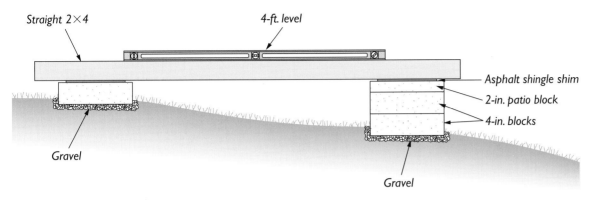

Straight 2×4   *4-ft. level*

Asphalt shingle shim
2-in. patio block
4-in. blocks

Gravel

Gravel

Leveling a Sloped Site
After laying out all the blocks for the shed foundation, use a long, straight 2×4 and a 4-ft. level to make sure each block is level with the others. If you need to raise a block just a little bit, place a thin shim of wood or asphalt shingle on top.

of the foundation so it is even with the highest corner.

If the building site is high and dry, you can set the blocks directly on the ground. However, if there's any chance that rain runoff will occasionally drain under the shed, you'll need to use a shovel to remove a patch of grass under each block, compact the soil with a hand tamper, then cover the exposed dirt with 2 in. or 3 in. of gravel before setting the blocks. The gravel bed will ensure that the soil beneath the blocks won't wash away or become soggy.

Precast pier blocks

This building method is similar to the solid-concrete block foundation discussed above. However, instead of using flat blocks, a series of precast concrete pier blocks are used to support the shed's floor frame. The pyramid-shaped blocks are designed for building decks, but they work great for sheds, too—provided you choose the right type.

There are a few styles of pier blocks available, including one that has a square hole molded into the top through which a vertical 4×4 post can be inserted. Another type has a flat wood block set

A series of precast pier blocks, arranged in three straight rows, provides a simple, secure way to support a floor frame. (Photo courtesy of DEKBRANDS™.)

A 2× joist fits into a slot molded in the top of this Dek-Block™ pier; the concrete pier will also accept a vertical 4×4 post. (Photo courtesy of DEKBRANDS.)

TRADE SECRET

Anchor bolts must be held in exactly the right spot until the concrete hardens. Suspend the bolts from small plywood hangers screwed to the top of the form (see below). Bore a hole near one end to hold the bolt.

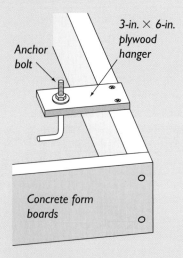

Anchor bolt

3-in. × 6-in. plywood hanger

Concrete form boards

IN DETAIL

Many do-it-yourselfers mistakenly use the words *concrete* and *cement* interchangeably. Although concrete is made with cement, the two materials aren't the same. Concrete is a blend of cement; sand; and an aggregate, such as gravel. Cement is a ground mixture of limestone, clay, shale, silica, and iron ore.

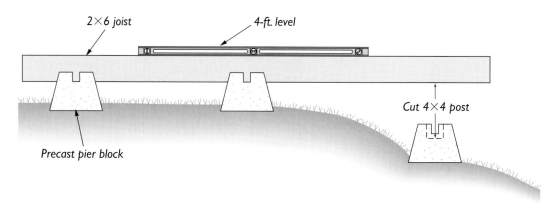

2×6 joist   4-ft. level

Cut 4×4 post

Precast pier block

Fitting a Post to a Pier Block

Dek-Block pier blocks can accept either a joist or a post, making them useful on very uneven sites. After setting a 2× joist into the slot of the pier that sits at the highest point, hold the joist level and measure the distance to the bottom of the socket in the lowest pier. Then cut a 4×4 post to fit.

into the top so you can toenail a joist in place. For building shed foundations, I prefer to use Dek-Block piers. Each block measures 8 in. high by 11 in. sq. and weighs about 45 lbs. Molded into the top surface are a 3½-in.-sq. recessed socket and a pair of 1½-in.-wide slots. The socket accepts a 4×4 post; the slots are used to support a 2× floor joist. Because Dek-Block piers can accept either a joist or a post, they can be used on very uneven sites and badly sloping terrain.

Skid foundations

When it comes to time-tested building methods, it's hard to beat a skid foundation. Builders have been using this type of on-grade foundation to support outbuildings for more than three centuries. The technique is surprisingly simple in both concept and application: Two or more long, straight timbers (skids) are laid on the ground in parallel, evenly spaced positions. The building's floor frame is then built on the skids, which are sometimes called runners or deadmen.

Skid foundations are still popular today, and it's easy to see why: They're very fast and easy to build; and they distribute the building's weight evenly over a broad surface. Unfortunately, because the timbers are long and straight, this type

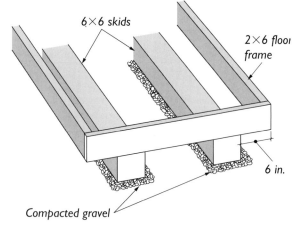

6×6 skids

2×6 floor frame

6 in.

Compacted gravel

Skid Foundation

A skid foundation is a simple and effective way to support the floor frame of a small shed.

of foundation is suitable only for sites that are relatively flat.

Originally, skids were nothing more than logs placed on the ground. Today, they're usually made of pressure-treated 4×6s, 6×6s, or 8×8s. You can also make skids by gang-nailing together three or four 2×6s or 2×8s and setting them on edge.

Although skids are often set directly on the ground, I prefer to lay them on a bed of gravel. The stone creates a very stable base that's not likely to settle or wash away. Begin by laying the skids in position on the ground, then mark around each one using spray paint or flour sprinkled from a

What Is Pressure-Treated Wood?

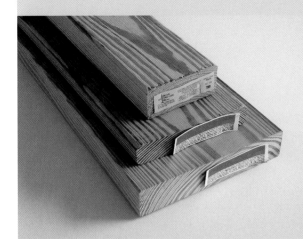

Treated lumber is labeled with information that includes chemical-retention levels and recommended uses. For foundations, use only lumber that's rated "Ground Contact." (Photo courtesy of SFPA.)

More than six billion board feet of pressure-treated (PT) lumber is sold annually, yet most people know very little about this ubiquitous green-tinted wood. Here are the facts.

Pressure-treated lumber is an exterior-grade wood that's chemically treated to resist rot, decay, and termites and other wood-boring insects. It's milled from softwood trees (typically southern yellow pine) and used for decks, fences, sheds, picnic tables, and playground equipment.

Waterborne chemical preservatives are pumped deep into the wood's fiber under extremely high pressure. Field tests conducted by the U.S. Forest Service have shown that treated lumber can resist decay and termite attack for more than 40 years. However, not all treated wood is created equal. The level of rot resistance is directly related to the amount of chemical preservatives in the wood. Lumber that's stamped "Above Ground Use" should be used only where it won't touch the ground, such as a deck railing or a fence board. Lumber that's designated "Ground Contact" can be used on or in the ground.

To make sure you're building with the right stuff, check the lumber's label or stamp for its "Chemical Retention Level." This number represents the minimum amount of preservative retained in the wood and is expressed in pounds of preservative per cubic foot of wood. The higher the number, the more rot resistant the wood is. The American Wood Preservers' Association has established the following retention levels for lumber treated with chromated copper arsenate:

Retention Level (lbs./cu. ft.)	Recommended Uses
.25	Above Ground
.40	Ground Contact
.60	Wood Foundation
2.50	In Salt Water

Lumber with a .40 retention level is adequate for building shed foundations, but for additional protection—and a little extra peace of mind—upgrade to a .60 retention level when building a pole-barn, timber-frame, or skid foundation.

For more than 60 years, the chemical most commonly used in treated lumber has been chromated copper arsenate (CCA). However, due to increasing health concerns, CCA-treated lumber will no longer be produced for residential projects after December 2003. It's predicted that lumberyards will run out of CCA-treated lumber by about April 2004. But don't worry, a new generation of treated lumber is already making its way into lumberyards. It's treated with an inorganic chemical (instead of arsenate) and is sold under various trade names, including Natural Select™, Preserve®, and NatureWood®. This new outdoor wood is often referred to as ACQ lumber, which stands for the chemicals alkaline copper quaternary.

PRO TIP

When mixing concrete in cold weather, use warm water to speed the curing process. If it's hot outside, mix in cold water to slow it down.

Skid Foundation on a Slope: Two Methods

If the site is out of level by more than 4 in. or so, it's easier to block up the low end of the skid than it is to excavate beneath the high end. The two best ways to do that are to make up the difference with solid-concrete blocks (A) or stack two skids on top of each other at the low end (B).

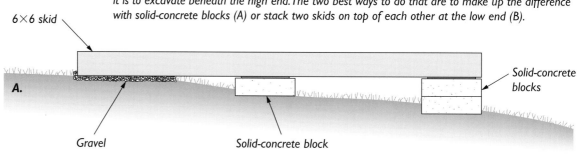

6×6 skid

A.

Gravel

Solid-concrete block

Solid-concrete blocks

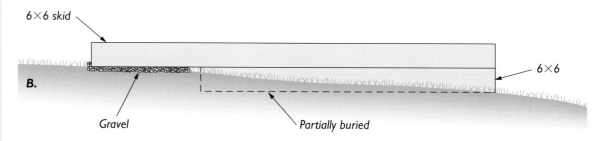

6×6 skid

B.

Gravel

Partially buried

6×6

TRADE SECRET

For extra protection against frost heave, take this step when building a frost-proof foundation. Wrap the wood posts or fiber-form tubes in thick plastic sheeting before setting them in the ground. Make sure the plastic extends all the way to the top of the holes. The frost won't be able to grasp onto the slippery plastic, thus greatly reducing the chances it'll upset the shed.

WHAT CAN GO WRONG

It doesn't seem possible that there's a wrong way to dig a hole, but there is when you're digging a hole for a concrete footing. First, don't make the hole any wider than necessary; you don't want to disturb the surrounding soil. More important, make the hole wider at the bottom than at the top to create a broad, stable base.

can. Move the skids out of the way, then use a flat shovel to remove the sod and about 2 in. of soil from the marked areas. Check the excavated areas to make sure they're close to being level. If they're not, remove a little more soil from the high spots. Next, add 3 in. to 4 in. of gravel. Compact the gravel with a hand tamper or gas-powered plate compactor, then replace the skids.

Timber-frame foundations

This foundation consists of little more than a rectangular wooden frame sitting on a gravel bed. The shed walls are built on the frame, and the entire weight of the building is transferred directly to the ground. However, the real advantage with this type of foundation is that you get to choose from a variety of flooring options. For example,

Hitting the Skids

If there's even the slightest chance that you may someday want to move your shed to another location, make the job easier by modifying the skids before you set them in place. Start by trimming off the bottom corners of the skids at a 45-degree angle so they'll slide more easily over the ground. Also, bore a 1½-in.-dia. hole about 4 in. from each end. That way, you'll have a convenient place to hook up a tow chain or steel cable.

Modify the ends of the skids, as shown here, if there's even the slightest chance that you may have to move the shed to another location.

I used brick pavers to create a handsome, hard-wearing floor for the Lean-to Shed Locker on p. 76. The floor area within the timber frame could also be filled with gravel, concrete, crushed granite, marble chips, or slabs of bluestone or slate. Another flooring option is to nail pressure-treated 2×6s over the frame in a manner similar to that of a deck. Just keep in mind that the 2×6s must be installed before you erect the shed walls.

A foundation frame is typically made from pressure-treated 4×4s, 4×6s, or 6×6s. The timbers are joined with half-lap corner joints or stacked two or three high and fastened together with long landscaping spikes or screws.

Frost-Proof Foundations

Frost-proof foundations extend deep into the ground to prevent freeze/thaw cycles from upsetting the building. They're generally required by code in cold-weather regions for sheds larger than 200 sq. ft. or taller than 12 ft. However, building codes differ from state to state, so be sure to check with the local building department for the exact requirements in your area.

Frost-proof foundations are typically more difficult to build than on-grade types, but how much harder depends on the soil. If the ground is soft and sandy, you may be able to excavate it with a

✓ According to Code

Generally speaking, you're not allowed to pour concrete for a frost-proof foundation until the building inspector has examined the holes. This is the building department's only way to know whether the holes are dug to the proper depth. Note that most inspectors are obliged to respond to your call for an inspection within a certain time period, usually 24 hours.

A) Half-lap Joint Construction

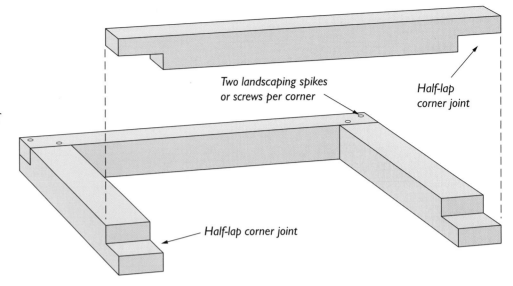

Two landscaping spikes or screws per corner

Half-lap corner joint

Half-lap corner joint

B) Stacked Corner Construction

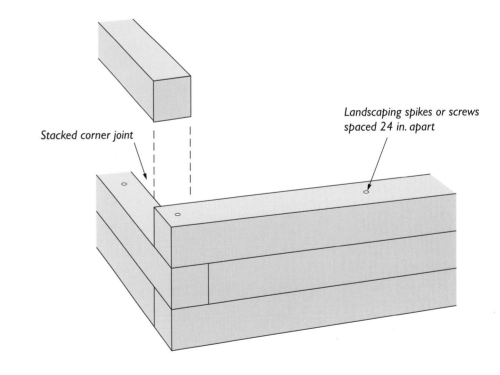

Stacked corner joint

Landscaping spikes or screws spaced 24 in. apart

Timber Frame Foundations: Two Options
A sturdy timber-frame foundation can be built with either half-lap joints (A) or stacked corners (B); it is useful when a gravel, stone, or brick floor is needed for the shed.

PRO TIP

When working with large timbers, swap your 16-oz. hammer for a 22-oz. to 26-oz. framing hammer, which has a longer handle and heavier head.

TRADE SECRET

Hardware can be attached to cured concrete with specialty masonry fasteners, which can make layout a little easier during a messy concrete pour. It's always best to avoid drilling into concrete if you don't have to, and screw-down brackets aren't nearly as strong as cast-in-place ones are. You'll find a variety of systems for fastening hardware to concrete at your local hardware or building supply store. I've found that some sort of an epoxy-encapsulated fastener works best, though such products are expensive and somewhat hard to find.

TRADE SECRET

Slate, bluestone, and unglazed clay tile are very porous and should be protected from stains with clear masonry sealer. The secret, though, is to brush on the sealer before setting the stone or tile in the concrete slab. Applying the sealer first makes it much easier to clean off any mortar or grout from the surface.

shovel and post-hole digger; if it's hard-packed clay or very rocky, you'll have to bring in a backhoe.

Poured-concrete piers

In its simplest form, a pier is nothing more than a column of concrete poured into a hole that extends below the frost line. Two or more rows of piers are used to support the shed's floor frame, similar to the way solid-concrete blocks are aligned for an on-grade foundation.

There is a wide variety of galvanized-metal framing hardware—such as post anchors, beam connectors, and tie-down straps—that can be used with poured-concrete piers. These specialty brackets provide a quick, easy way to create strong, lasting joints between the concrete piers and the wooden joists, posts, or carrying beams.

Calculating Concrete

To figure out how much concrete you'll need for pouring a slab or footing, refer to the chart below or use the following formula: Multiply the length times the width times the thickness of the slab, then divide by 12. Divide again by .45 for the number of 60-lb. bags of ready-mix concrete you'll need. If you're using 80-lb. bags, divide by .60 for the number of bags needed.

4-in.-Thick Slab (total sq. ft.)	Number of 60-lb. Bags of Concrete Needed	Number of 80-lb. Bags of Concrete Needed
6×8 ft. (48 sq. ft.)	36	27
8×10 ft. (80 sq. ft.)	59	44
10×12 ft. (120 sq. ft.)	89	67
12×14 ft. (168 sq. ft.)	124	93

To estimate how much concrete you'll need for the piers, use the chart below. Note that there's information for both 60-lb. and 80-lb. concrete bags and for holes ranging in depth from 24 in. to 42 in. and in diameter from 8 in. to 18 in.

Number of Bags of Concrete Needed

	60-lb. Bags				80-lb. Bags			
	Diameter of Hole				Diameter of Hole			
Depth of Hole	8 in.	10 in.	12 in.	18 in.	8 in.	10 in.	12 in.	18 in.
24 in.	1.5	2.5	3.5	8.0	1.25	2.0	2.66	6.0
30 in.	2.0	3.0	4.5	10.0	1.5	2.5	3.33	7.5
36 in.	2.33	3.6	5.25	12.0	1.75	3.0	4.0	9.0
42 in.	2.75	4.25	6.25	13.75	2.0	3.25	4.6	10.5

Poured-Concrete Piers: Three Methods
There are many ways to pour a pier foundation. Here are three techniques that work well for virtually any size outbuilding:

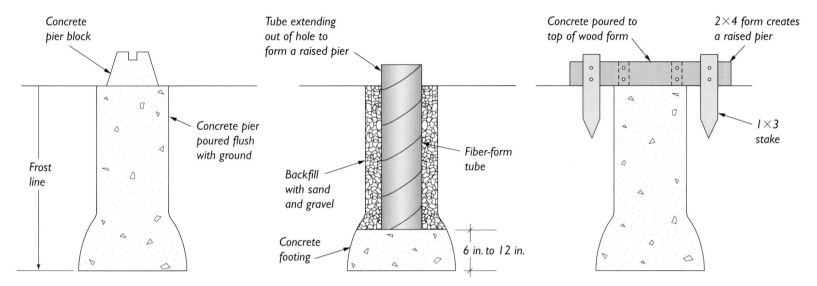

Solid Pier with Pier Block
Dig 12-in.-dia. holes down to the frost line, then pour concrete piers flush with the ground. Set precast concrete pier blocks or solid-concrete blocks on top. The blocks can be set in the wet concrete or simply laid on top once it cures. If you set them in the wet concrete, make sure all the blocks are level before the concrete hardens.

Sonotube Pier
Pour a 6-in. to 12-in. pad of concrete —called a footing—into the bottom of the hole and let it cure overnight. Next, stand a round fiber-form tube (commonly known by the trade name Sonotube) on top of the footing. Backfill around the tube with soil, then fill it with concrete. The advantage of this method is that before installing the tube, you can cut it flush with the ground or let it protrude above the hole to create a raised pier.

Formed Pier
Build a square form out of 2× lumber and set it over the center of the hole. Hold the form in place with a couple of 1×3 stakes pounded into the ground. Then fill the hole with concrete, bringing it right to the top of the form. After the concrete cures, strip away the form to reveal a raised pier.

Plan View of Form

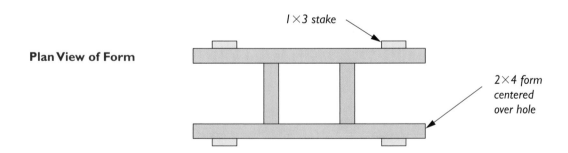

Just make sure you set the brackets in the piers before the concrete hardens.

Poured-concrete slab

The average backyard storage shed doesn't really need a poured-concrete floor, but it's the best choice for large outbuildings that will be used to store heavy equipment, such as woodworking machines, tractors, boats, motorcycles, snowmobiles, and antique cars.

There are two basic ways to pour a concrete floor, but only one qualifies as a frost-proof foundation. It's called a monolithic slab, because the floor and the perimeter foundation walls are all

PRO TIP

When using a circular saw to cut plywood, lay the sheet with the best face down. That way, any splintering—called tearout—will occur on the top, or back, surface.

WHAT CAN GO WRONG

One of the most common and chronic of all carpentry mistakes occurs when framing a rough opening in a wall for a window or door. The problem is caused by the somewhat confusing way that manufacturers and retailers specify the size of a window or door. For example, a 48-in.-wide window is sold as a *4-0* unit (pronounced *four-oh*). A *3-8* door isn't 38 in. wide; it's 44 in. wide. If you're not paying close attention, it's very easy to frame a 30- in.-wide opening for a 3-0 door, then discover that the opening is 6 in. too narrow. Avoid these costly, time-consuming mistakes by double-checking the dimensions on the plans before ordering windows and doors. Also, place your order early and have the units on hand before starting construction. That way, you can measure each window and door before framing the rough openings.

Is Burying Posts a Good Idea?

There's some debate about whether or not you should bury a wood pole—even a treated one—in concrete, where it may eventually rot. Some builders prefer to pour raised concrete piers and set the poles on top. (See the drawing on the facing page.) This method does help the poles last longer, but much of the structural integrity of a pole-barn foundation comes from the fact that the poles extend deep into the ground. Raising them out of the ground can weaken the structure.

Ordinarily, I don't like to bury wood in the ground, but it's definitely the best way to gain the strength needed when building a pole barn or setting a gate post or an end-of-the-run fence post. I've buried many pressure-treated posts over the years and have always questioned the process, but on the other hand, I've never had to replace any of them. I spoke with a local pole-barn builder who has been burying wood—pressure-treated and creosote-soaked—for more than 30 years and continues the practice to this day with confidence.

poured at the same time. The walls extend down to the frost line and are usually between 12 in. and 16 in. thick. The floor itself is only 4 in. to 6 in. thick, but it's reinforced with wire mesh or metal reinforcing bars.

The second type of concrete floor is known as a floating slab or an on-grade slab. It's nothing more than a 4-in. to 6-in. layer of concrete sitting on the ground. This type of floor should never be used when the plans or local building codes call for a frost-proof foundation.

Pole-barn foundations

All the foundations discussed so far are designed to support the floor of an outbuilding. The floor, in turn, supports the walls. A pole barn is completely different. In fact, it technically doesn't even have a floor.

Pole-barn construction starts with a series of holes dug below the frost line around the perimeter of the foundation. Concrete footings are poured into the bottom of each hole, then tall, decay-resistant round poles or square timbers, which extend all the way to the tops of the walls, are set in the holes. Horizontal beams are bolted along the tops and bottoms of the poles to tie

everything together and support the walls and roof framing. (See the bottom left photo on p. 176.)

Instead of a wood-framed floor, a pole barn's floor is actually the ground. To keep it from wearing away or turning muddy, the area is covered with several inches of processed stone, pea gravel, or wood chips. As a result, the floor is basically flush with the surrounding grade, making a pole-barn foundation perfect for outbuildings that house lawn tractors, boat trailers, farm machinery, horses, and livestock.

As with other types of foundations, there are several ways to build a pole barn. For the Gambrel

+ SAFETY FIRST

Pressure-treated wood poses no health hazards if you follow a few simple precautions. Be sure to wear gloves when handling treated lumber, and wash up thoroughly before eating or drinking. Always wear safety goggles and a dust mask when cutting or drilling treated wood, and be sure to properly dispose of all scraps; never burn treated wood.

Pole-Barn Foundations: Three Options

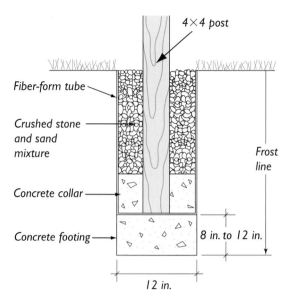

4×4 post

Fiber-form tube

Crushed stone and sand mixture

Concrete collar

Concrete footing

Frost line

8 in. to 12 in.

12 in.

Post in Sonotube

This basic pole-barn foundation is strong and simple to build; the poured concrete footing creates a solid, frost-proof base, while the tube and stone mixture leave little for frost to grab onto.

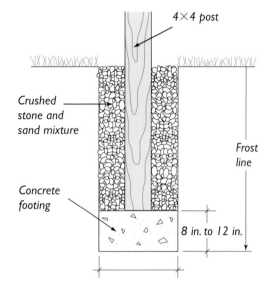

4×4 post

Crushed stone and sand mixture

Concrete footing

Frost line

8 in. to 12 in.

The diameter of the hole is three times wider than that of the post.

Post on Footing

A simpler option is to eliminate the tube and pour the concrete footing directly into the bottom of the hole. Then after placing the vertical post on top of the footing and centered in the hole, backfill the entire hole with a mix of crushed stone and sand.

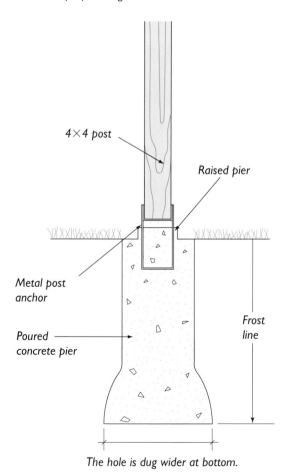

4×4 post

Raised pier

Metal post anchor

Poured concrete pier

Frost line

The hole is dug wider at bottom.

Post on Raised Pier

Raising the pier above ground level by pouring a concrete pier avoids the problem of rot, but it may weaken the strength of a pole-barn foundation.

PRO TIP

A power miter saw won't replace your portable circular saw, but it will offer a quicker, more accurate way to crosscut framing lumber.

IN DETAIL

I prefer using tongue-and-groove plywood for shed floors, even though it costs more and is more difficult to install than standard square-edged plywood. The reason? When the plywood sheets are butted together, the tongue on one sheet locks tightly into the groove on the adjoining sheet. The result is a strong, rigid seam that won't sag, even where it spans open spaces between joists.

Shed walls are typically framed with 2×4s. Save time by building the walls on the ground, then tip them up into place.

Storage Barn (see p. 168), fiber-form tubes were placed in the holes, then an 8-in.-thick concrete footing was poured into the bottom of each tube. After the concrete cured, 4×4 pressure-treated posts were placed on top of the footings and another 12 in. or so of concrete was poured in. The rest of the tube was then filled to the top with a mixture of crushed stone and sand. I like this technique because the poured concrete creates a solid, frost-proof base and the sand and stone mixture leaves little for frost to adhere to.

+ SAFETY FIRST

Be aware that a construction site is a potentially dangerous place for children and pets. Never allow kids to play near lumber piles; boards could shift and collapse, pinning someone beneath them. Cover footing holes with plywood to prevent pets and toddlers from falling in. Finally, put away all your tools at the end of each day.

Wall-Framing Techniques

There are three common construction techniques used to build shed walls. A vast majority of outbuildings—probably more than 90 percent—are stick-built out of 2×4s. To be honest, it's hard to justify using any other technique, especially if you're building a basic storage shed. However, if you're building a workshop, writer's studio, or home office, you may want to put up a post-and-beam building. The exposed timbers lend a more finished, handcrafted look to the interior. This is also a more interesting and challenging way to build.

A pole-barn building, as discussed previously, is suspended off the ground by several round poles or square posts. The wall framing is fastened to the poles and can be made out of large timbers or 2× lumber. The exact wall-framing technique depends on the type of poles or posts used and your choice of siding material.

Stick-built construction basics

It's no mystery why carpenters prefer stick-built construction: It's the fastest, easiest, and most affordable way to frame walls. The term *stick-built* refers to the fact that the walls are constructed out of individual "sticks" of lumber: typically 2×4s or 2×6s. This technique is also called western or platform framing.

Headers are an important component in stick-built construction and are used to transfer roof loads above doors and windows down through the framing and onto the foundation. Their size and thickness depend on several factors, including the width of the opening, the height of the building, and whether the wall is a load-bearing one. If you're not sure how big to make the headers, ask the building inspector when you submit your plans and apply for a building permit.

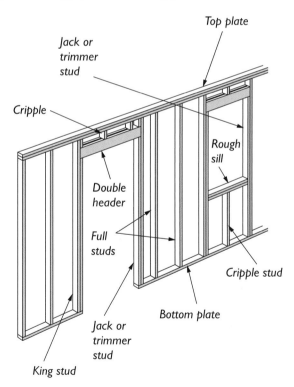

Stick-Built Wall
Each section of a stick-built wall consists of horizontal top and bottom wall plates and vertical studs, which are usually spaced 16 in. on center. Horizontal beams, called headers, run across the top of each window and door opening and transfer weight from above the window or door to the jack and king studs on each side of the opening.

Nominal Numbers

Framing lumber is referred to by its "nominal" dimension, not by its actual dimension. That's because a log is ripped into 2×4s during its initial pass through the sawmill, the rough sawn boards measure a full 2 in. by 4 in. However, after the final milling stage and drying process, they're only 1½ in. thick by 3½ in. wide. In fact, all 2× lumber is actually 1½ in. thick, and all 1× boards are actually ¾ in. thick.

Use the chart below as a handy reference for both nominal and actual dimensions of standard lumber sizes.

Nominal Dimension	Actual Dimension	Nominal Dimension	Actual Dimension
2×3	1½ × 2½	2×12	1½ × 11¼
2×4	1½ × 3½	4×4	3½ × 3½
2×6	1½ × 5½	4×6	3½ × 5½
2×8	1½ × 7¼	6×6	5½ × 5½
2×10	1½ × 9¼	8×8	7¼ × 7¼

You can use standard framing lumber for building walls; Douglas fir, hemlock, and spruce are fine. However, it's important to use pressure-treated lumber for any framing member, such as the mudsill, that comes in contact with the ground or a concrete foundation.

Post-and-beam construction

This ancient building method uses large vertical posts and horizontal beams to form the skeletal frame of the walls. There are far fewer parts in a

The frame of a post-and-beam building has far fewer parts than does a stick-built shed, but the parts are also much heavier. (Photo © Smith-Baer.)

PRO TIP

Save time and reduce costly measuring errors by cutting a pair of rafters, checking their fit, and then using them as a template to mark the rest.

IN DETAIL

Building plans often refer to *on center* dimensions; for example, rafters are spaced 24 in. on center. That means the distance from the center of one board to the center of the next measures 24 in., not that there's a 24-in. space between them.

WHAT CAN GO WRONG

When installing roof rafters or prebuilt trusses, it's important to place each one directly over a wall stud. That way, the weight of the roof will be transferred directly to the foundation. If you set the rafters or trusses between the studs, the weight can bow the plates or eventually crush the walls.

Plywood Veneer Grades

The wood veneers used in the manufacture of plywood are graded according to a set of quality standards established by the American Plywood Association. A letter grade is assigned to the veneer that's bonded to the front and back surfaces of each plywood sheet. Here are brief descriptions of the five most common veneer grades:

A	The best veneer grade. Has a sanded, smooth surface with less than 18 neatly made repairs. Defects are concealed with synthetic filler or wood patches.
B	Has a sanded, solid surface. Tight knots up to 1 in. in diameter are permitted. Some minor splits are also allowed.
C	Has tight knots up to 1½ in. in diameter and knotholes up to 1 in. in diameter. Discoloration and sanding defects are permitted. Limited splits and stitched repairs are also allowed.
C-Plugged	An improved C-grade veneer with a plugged and sanded surface. Splits up to ⅛ in. wide are permitted and knotholes up to ¼ in. by ½ in. Some broken grain is also allowed.
D	Contains knots and knotholes up to 2½ in. in diameter. Limited splits and stitched repairs are permitted.

post-and-beam wall compared to those in a stick-built wall, but the parts are larger, so they're a little more difficult to work with. The posts and beams are usually cut from 4×6s, 6×6s, or 8×8s.

There are various ways to join the posts to the beams, but the traditional method of using mortise-and-tenon joinery is still preferred today.

Fit the beam's tenon into the mortise in the post. Hold the joint closed with long wooden dowel pegs or galvanized screws. (Photo © Smith-Baer.)

Each joint consists of a recessed slot (mortise) and a protruding tab (tenon). Mortises are cut by first drilling out the waste wood with a large-diameter drill bit, then squaring up the hole with a mallet and chisel. Tenons are typically cut with a portable circular saw and a handsaw. When a tenon fits precisely into the mortise, the two parts form a strong, lasting joint.

A post-and-beam structure can be covered with plywood sheathing and siding like any other building. Or the siding can be set within the framework of the building so the posts and beams will remain visible from both inside and out. (See the photo on p. 13.)

Pole-barn framing

The method used to frame the walls of a pole barn depends largely on whether it has round poles or square posts. For round poles, it's easiest to nail 2×4s or 2×6s horizontally across them. If the walls will be covered with plywood siding, three boards should suffice. If you're planning to install board-and-batten siding or some other vertical-board exterior, you'll need to install four or more equally spaced 2× boards to provide adequate nailing for the siding.

If the building has square posts, you have two options: Nail 2×4s or 2×6s across the posts, as described above for round poles, or build framed sections out of 2×4s and set them between the posts.

Shed Floors

The building material used for the floor of a shed is often determined by the foundation. For instance, a wood-joist floor frame is usually covered with plywood. However, a timber- frame foundation can have a floor made of brick, gravel, or wood. Also, consider how the outbuilding will be used before choosing the type of floor. It wouldn't make sense to install carpeting in a tool-storage shed, but it might be a good choice for a kid's playhouse or a writer's studio.

Plywood floor

Plywood is an excellent flooring choice for most stick-built outbuildings. It's easy to install, affordable, and surprisingly strong. Plus, the large sheets create a very smooth, flat surface, which makes it easy to roll, drag, and push objects across the floor. However, there are dozens of types and thicknesses of plywood. If you install the wrong one, the floor will sag, crack, and eventually rot right out from under you.

Fortunately, picking the right plywood for a shed floor is easy. Just follow this simple rule: Use only ¾-in.-thick, exterior-grade plywood. Anything thinner—even ⅝-in.-thick plywood—won't provide the necessary support. Although you could build a floor with two layers of ½-in. plywood that would be slightly stiffer than one

Pole Barn Wall Framing: Two Options
How the walls of a pole barn are framed depends largely on whether the posts are round (A) or square (B). Some 2×4 nailers for plywood or vertical siding can be nailed horizontally to either round or square posts; conventionally framed wall sections can be set between square posts.

A) Round-Pole Wall Framing

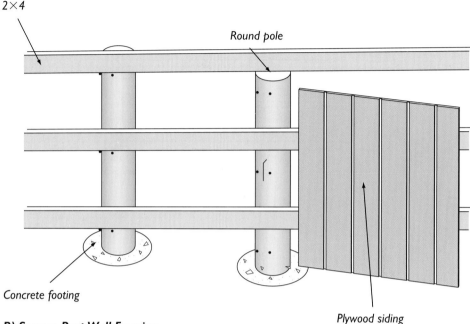

2×4

Round pole

Concrete footing

Plywood siding

B) Square-Post Wall Framing

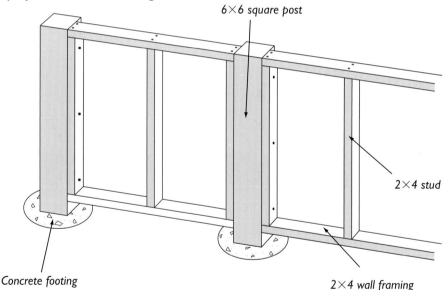

6×6 square post

2×4 stud

Concrete footing

2×4 wall framing

layer of ¾-in. plywood, the added strength wouldn't be nearly enough to justify the added expense and work.

Standard plywood is manufactured with ordinary glue and should never be used for a shed floor. Its thin wood layers (plies) will start to delaminate and blister in less than a year. Use only

PRO TIP

While a roof can be framed conventionally with individually installed rafters, a faster and safer way to frame a roof is with simple site-built trusses.

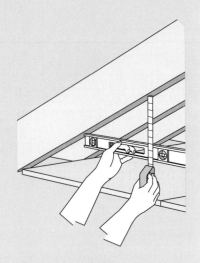

Measuring Roof Slope

TRADE SECRET

Here's how to determine the slope of a roof—without climbing onto it. Make a mark 12 in. from one end of a level. Place the level against the bottom of a roof rafter and hold it perfectly level. Next, measure vertically from the 12-in. mark straight up to the underside of the rafter. That measurement is the number of inches the roof rises in 12 in. If the distance is 6 in., for example, then the roof has a 6-in-12 slope.

The typical shed floor consists of a 2×6 frame topped with ¾-in.-thick plywood. The 2×6 joists are spaced 16 in. on center.

exterior-grade plywood for the floor. Its plies are glued together with an all-weather, water-resistant adhesive.

I prefer to use ¾-in. ACX plywood for shed floors. The "A" and "C" designations refer to the grades of veneer used on the front and back surfaces. The "X" identifies it as an exterior-grade product. You can save a few dollars by installing BCX and CDX plywood instead, but they're not as smooth as the A-grade sheets.

The plywood is fastened to the floor joists with 1⅝-in. decking screws or 2-in. (6d) galvanized

Use decking screws or ring-shank nails spaced 8 in. to 10 in. apart to attach the plywood decking to the floor frame.

ring-shank nails. Most plywood floors are left bare, but they'll last longer and clean up easier if you apply two coats of enamel deck paint. If the building will be used as an office or studio, consider covering the plywood with resilient vinyl flooring, carpeting, laminate planks, or prefinished wood strips.

Wood-plank floor

An alternative to a plywood floor is traditional solid wood planking. This type of flooring consists of fat tongue-and-groove planks that measure 1½ in. to 2 in. thick and 6 in. to 8 in. wide.

Although a solid wood floor is considerably more expensive than a plywood one and takes much longer to install, it does have a couple of distinct advantages. First, a wood-plank floor just looks great, instantly lending old-world charm to an interior space. Thick planks also create a stable, rock-solid floor that won't bounce or bend, even under extreme weight. This type of flooring is a good choice for woodshops that house heavy machines and large workbenches.

The planks are typically installed perpendicular to the floor joists, but you can also lay them diagonally for a more interesting appearance.

A timber-frame foundation provides a chance to lay a beautiful brick floor. Set the bricks in a bed of gravel or sand. (Photo © Smith-Baer.)

Just be aware that a diagonal pattern creates 10 to 15 percent more cutoff waste.

Dry-laid brick floor

The main reason I enjoy building a timber-frame foundation is that it affords me the opportunity to put in a brick floor. This style of shed floor is attractive, extremely durable, and fun to install. The installation technique I use is called a dry-laid method because the bricks are simply set down on a bed of sand or crushed stone; they're not adhered with mortar or mastic.

For this type of floor, it's important to use either concrete brick pavers or hardened clay bricks that are specifically designed for use on walkways and patios. Don't use standard clay wall bricks; they're too soft and porous. I'll explain how to install this type of floor in more detail in Chapter 4.

The dry-laid method can be used to install a variety of masonry materials, including cut granite, bluestone, patio blocks, and flagstone. An easier, though less attractive, alternative to laying bricks is to completely fill the frame with concrete, crushed stone, pea gravel, marble chips, or some other loose-fill masonry material.

Slate-on-slab floor

There are few shed floors as strong or as durable as a poured-concrete slab. But let's face it, concrete isn't particularly attractive. One way to dramatically improve the look of a slab floor is to cover it with slate. Again, this probably isn't something you'd do to the floor of a tool-storage building, but it would be terrific for a potting shed or a pool cabana.

Slate is an extremely hard, natural stone that comes in various shades of gray, green, blue, and dark red. It's commonly available in square and rectangular tiles and irregularly shaped slabs. The machine-cut tiles are about ¼ in. thick; slate slabs

range in thickness from about ¾ in. to 1¼ in. thick. Both styles are adhered to a fully cured slab with mortar, but the techniques differ slightly.

Because slate tiles are uniformly sized, they can be installed with thin-set mortar. Buy the latex-fortified type and mix it with water. If you can find only standard, unadulterated thin-set, then pick up a jug of liquid latex additive and use it, instead of water, to mix up the mortar. The latex will increase the mortar's bond strength and make it much more water-resistant. Slate slabs can't be installed in thin-set because their surfaces are too irregular. They must be laid in a thick bed of standard masonry mortar. This method is referred to as a mud job because the mortar is so thick.

Roof Framing

A few rafters and a ridge board are all you need to frame nearly any shed roof—and in some cases, you don't even need the ridge board. Here, we'll take a look at five styles of sloped roofs: gable, shed, saltbox, gambrel, and hip.

Gable and shed-style roofs are the simplest to build, but many folks find saltbox and gambrel

Speed the job of framing the roof by prefabricating the roof trusses on the ground. Then lift each truss into place and screw it to the walls.

PRO TIP

When installing the ledger for a shed roof, be sure to fasten it to solid framing inside the wall, not just to the siding or plywood sheathing.

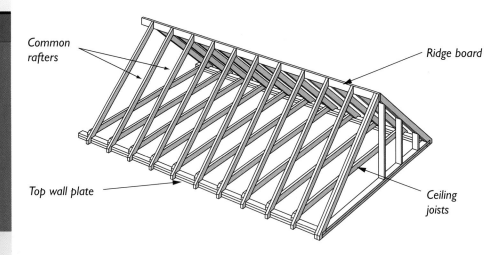

Common rafters

Ridge board

Top wall plate

Ceiling joists

Gable-Roof Framing
A gable roof's familiar A-shape profile is formed by pairs of common rafters that run at an angle from the tops of the walls to the peak.

TRADE SECRET

It's possible to build an entire gambrel roof without cutting a single miter joint. Special metal framing brackets are formed at the precise angles needed to create the double-sloping roof. All you need to do is insert square-cut 2×4s into the brackets and secure them with screws; there are no angles to figure out or gusset plates to install. This shortcut will allow you to frame a gambrel roof in half the time it may ordinarily take. The brackets are available through various mail-order catalogs and at specialty hardware stores. The Fast Framer™ set shown here is available from Lee Valley Tools®.

These metal framing brackets allow you to build an entire gambrel roof without cutting a single miter joint. (Photo courtesy of Lee Valley Tools.)

roofs more interesting. Hip roofs have a distinctive, refined appearance, but they're rather tricky to build, which may explain why you don't see them on very many outbuildings.

The traditional way to frame a roof is to cut and install the rafters, ridge board, and ceiling

Roof Slope vs. Roof Pitch

Technically speaking, the angle of a roof is known as the roof slope. It's calculated by the number of inches it rises vertically for every foot of horizontal run. For instance, a roof with a 10-in-12 slope rises 10 in. for every 12 in. of horizontal run. Most sheds have a roof slope ranging between 6-in-12 and 12-in-12.

The term *roof slope* is sometimes confused with *roof pitch,* which is an older, less-common phrase. Roof pitch is based on the roof's vertical rise divided by its span, or width; it's always expressed as a fraction. For example, if a roof measures 8 ft. in height from the eave to the ridge, then it has an 8-ft. rise. If a shed is 15 ft. wide with a 6-in. overhanging eave on each side, it has an overall span of 16 ft. In that case, the roof pitch is 8 over 16, or ½.

joists one board at a time. The advantage of this piecemeal approach is that a single person working alone can easily lift and nail the boards in place. The drawback is that you spend a lot of time clambering up and down ladders, which is tiring and potentially dangerous.

A second roof-framing option is to assemble the roof trusses on the ground, then raise them into place. This technique—used for the sheds in this book—is easier and safer than working from a ladder, but you'll need two or three people to lift the trusses onto the walls.

Gable roof

A gable roof has two sloping planes of equal length. Its familiar A-shaped profile is formed by pairs of common rafters that run at an angle from the tops of the walls up to the peak. A ridge board, if used, runs horizontally between the pairs of rafters where they meet at the peak.

The size and spacing of the lumber used to frame a roof varies, depending on the size of the building. The bigger the building, the larger the boards must be. Most storage sheds are framed with 2×4 or 2×6 rafters and joists, which are spaced 16 in. or 24 in. on center. The ridge board is usually cut from a continuous length of 1×6 or 1×8.

The gable roof on a post-and-beam building is framed with larger timbers, such as 4×4s and

4×6s, but fewer rafters are needed because the hefty framing can support more weight.

When framing a roof with site-built trusses, you don't need a ridge board. Just make sure the rafters are properly positioned before nailing on the plywood sheathing. (For detailed information about building a gable roof with trusses, see the Colonial-Style Shed on p. 152.)

Shed roof

A shed roof is basically half of a gable roof. In other words, it's a single sloping roof plane. You typically see this type of roof on small, narrow structures. Shed roofs can be used on freestanding outbuildings, but they're most commonly found on sheds attached to another structure, such as a house or garage. In that case, a horizontal board called a ledger is fastened to the wall to support the upper ends of the rafters.

Shed roofs are generally built at a shallower pitch than that of gable and saltbox roofs, with a roof slope ranging between 4-in-12 and 8-in-12. When deciding on the angle of a shed roof, take into account where the doors are located. If they're hung on the shed's sidewall, directly under the low edge of the roof, then consider a shallower roof slope. A steeply pitched roof will come down lower, reducing the headroom in the doorway. The only ways to prevent a steep roof from cutting into the doorway are to build taller walls or to make

The gable roof of this shed was framed with site-built roof trusses; note that there's no ridge board running along the peak.

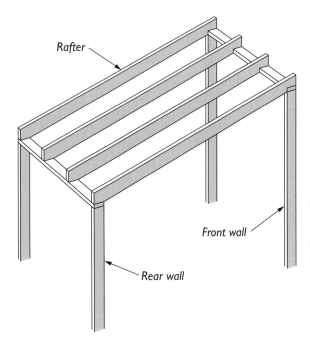

Rafter

Front wall

Rear wall

Shed-Roof Framing
Basically half of a gable roof, a shed roof is simply a single sloping roof plane. It is found on both attached and freestanding buildings.

✓ According to Code

If you're building in an area that receives significant snowfall, the shed's roof must satisfy a specific snow-load rating. This building code ensures that the roof is capable of supporting a certain number of pounds of snow per square foot. To find out whether your shed roof meets the requirements in your area, contact the local building inspector.

PRO TIP

Construct wooden steps, platform decks, and ramps from pressure-treated lumber rated for ground-contact use.

IN DETAIL

Some sheds are low enough to the ground that you don't need to build a step or a ramp. However, you should still create a landing in front of the door to keep the ground from wearing away and turning muddy. Start by building a rectangular wooden frame from pressure-treated 2×4s. Make it slightly longer than the doorway and at least 24 in. wide. Dig out the soil in front of the shed and set the frame flush with the ground, then fill the space within the frame with gravel or marble chips.

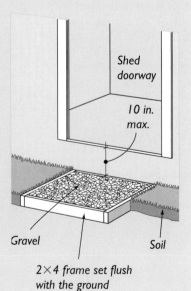

Shed doorway

10 in. max.

Gravel

Soil

2×4 frame set flush with the ground

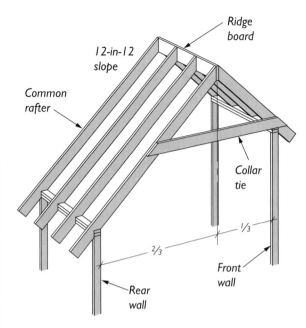

Ridge board

12-in-12 slope

Common rafter

Collar tie

1/3

2/3

Front wall

Rear wall

Saltbox-Roof Framing
Like a gable roof, a saltbox roof is framed with pairs of rafters that meet at the peak, though one roof plane is slightly longer than the other.

the entire shed narrower—not very practical solutions. (To learn more about building a shed roof, see the Lean-To Shed Locker on p. 76).

Saltbox roof

A saltbox roof is similar to a standard gable roof; both are framed with pairs of rafters that meet at the peak. The main difference is that a saltbox has one roof plane that's slightly longer than the other. This single design change shifts the roof peak off center so it's closer to the front wall, thus creating the distinctive saltbox roof.

Unfortunately, traditional saltbox design has been altered, distorted, and modernized so often that it's sometimes hard to recognize. Today's builders have tried to "improve" the design by changing the angle of the roof slope, raising the walls, and shortening the long roof plane. The result is an ugly, out-of-proportion building. To ensure that your saltbox roof remains true to its

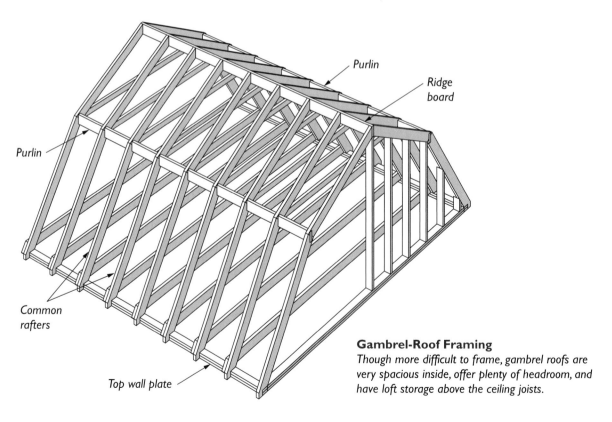

Purlin

Ridge board

Purlin

Common rafters

Top wall plate

Gambrel-Roof Framing
Though more difficult to frame, gambrel roofs are very spacious inside, offer plenty of headroom, and have loft storage above the ceiling joists.

When framing a large gambrel roof, it's easier to build trusses on the ground, then lift them into place.

colonial roots, follow these two essential design rules. First, frame the roof to a 12-in-12 slope (that's 45 degrees). Second, create the proper proportions by situating the roof peak one-third of the way back from the front wall. (To learn more about building a saltbox roof, see the Potting Shed on p. 106.)

Gambrel roof

The classic gambrel roof is easily recognized by its distinctive double-sloping profile. This traditional barn-style design features two short, shallow roof planes that come off the ridge and then break sharply down to longer, steeper slopes.

Gambrel roofs are a bit more difficult to frame than gable, shed, or saltbox roofs, mainly because they contain many more parts. However, gambrel roofs are very spacious inside, offer plenty of headroom, and provide loft storage above the ceiling joists. Plus, this style of roof can be built for storage buildings of all sizes.

If you're thinking about building a gambrel roof, remember that the doors must be hung on the end walls because the shorter sidewalls aren't tall enough. Keep this in mind when laying out the shed's foundation, making sure the end wall with the doors faces in the desired direction.

Like most other roof styles, a gambrel roof can be framed piecemeal, one board at a time, or you can assemble trusses on the ground and then lift them into place. (For step-by-step instructions on framing a gambrel roof, see the Gambrel Storage Barn on p. 168.)

Hip roof

A hip roof is essentially a gable roof with four sloping planes. Two planes slant down from the ridge to the sidewalls (as with a standard gable) and one plane at each end slopes down to the top of each end wall. This quadruple-slope design creates an overhanging eave that wraps around the entire building, a feature that's unique to hip roofs.

Although hip roofs are distinctive looking, you don't see them very often on outbuildings. First, they're much more difficult and time-consuming to frame than other roof types. That's because a hip roof has three kinds of rafters: common, hip,

TRADE SECRET

Wooden ramps and steps can become slippery, especially when they're wet. To reduce the likelihood of slipping and taking a nasty fall, apply strips of nonslip abrasive tape to the treads. The adhesive-backed tape comes in various widths and is sold by the linear foot at most home centers and hardware stores.

IN DETAIL

I wish I could give you several practical reasons why you should build a saltbox roof for your storage shed, but there really aren't any. In fact, in some ways, this type of roof is impractical: The longer roof plane provides limited headroom, and the rear wall is too short for hanging tools. However, occasionally form wins out over function. The reason to build a saltbox is that there's no better way to capture the beautiful, elegant design of classic colonial architecture.

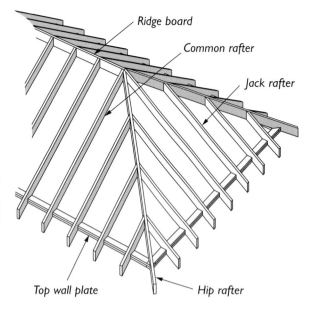

Hip-Roof Framing
While hip roofs are distinctive looking, they're difficult to frame. You don't see them very often on outbuildings.

and jack. To cut precise miter and bevel joints in these boards is challenging for even an experienced carpenter. Second, the hip portions slice through the "attic" area and eliminate an enormous amount of overhead storage space. Finally, a hip roof exposes a considerable amount of roof shingles to view. Most folks prefer to see a little more wood siding and a little less asphalt roof shingles.

Stair and Ramp Construction

If the floor of your shed is more than 8 in. or so off the ground, you'll need to build some sort of step or ramp. Keep in mind that you'll often be carrying or wheeling an item in or out of the shed. If the step down is too great, you could easily trip and injure yourself. A properly designed and sturdily constructed set of stairs or a ramp will make entering and exiting your shed safer and easier.

Two Ways to Build Steps

Notched Stringer

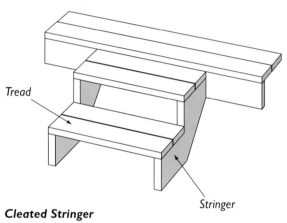

Cleated Stringer

When designing a set of steps, make the treads at least 10 in. deep and between 7 in. and 8 in. high. It's also very important to make sure all the steps are exactly the same height. Even the slightest discrepancy will create a tripping hazard.

Wooden steps

A set of wooden steps is the best choice when you need to step up two or three times to reach the shed's floor. One option is to build a short set of stairs with a pair of stringers and two or three treads in the traditional manner. Stringers can be cut from 2×12 stock, with two 2×6s used to make each tread. Stairs are usually built a few inches wider than the doorway opening, but in no case should they be less than 36 in. wide.

There are two basic ways to assemble stairs. Stringers can be notched to accept the treads, or the treads can rest on cleats nailed to the inside of the stringers. An alternative is to build a couple of simple wooden platforms and stack them on top of each other. Make the rectangular base of each platform from pressure-treated 2×6s, then cover it with 2×4s or 2×6s spaced about ¼ in. apart. Build the bottom platform at least 10 in. wider than the top one to create the first step. To prevent the platforms from sliding out of position, screw them together and to the shed.

Platform deck

A platform deck is similar to platform steps, but much bigger. The idea is to build two or three large wooden platforms of progressively smaller sizes and stack them on top of one another at the entrance to the shed. Again, build the frames out of pressure-treated 2×6s, with joists every 16 in., then fasten down 2×4 or 2×6 decking. For a more visually interesting effect, run the deck boards diagonally in alternating directions from one platform to the next.

Platform steps are easier to build and more attractive than a typical stringer-and-tread staircase. Set the platforms on solid-concrete blocks to keep them level and stable.

A two-tier wooden platform creates a warm and welcoming entrance to this workshop. The platform is part staircase, part sun deck. (Photo © Joseph Truini.)

PRO TIP

Mount an exterior light fixture to the shed, and install three-way switches at the house and shed. The light can then be turned on and off from either location.

IN DETAIL

One way to bring electricity to a shed—particularly if it is located in a relatively remote area—is by installing a solar-powered panel. The panel contains several photovoltaic cells that capture the sun's energy, which is then stored in a rechargeable battery. Keep in mind that a small solar panel like the one shown here can produce only enough power for a couple of lighting fixtures. You won't be able to run machinery.

Gain electrical power without wiring by using a roof-mounted solar panel. Photovoltaic cells inside the panel capture the sun's energy and store it in a rechargeable battery. The panel provides enough electricity to power a couple of light fixtures. (Photo © Joseph Truini.)

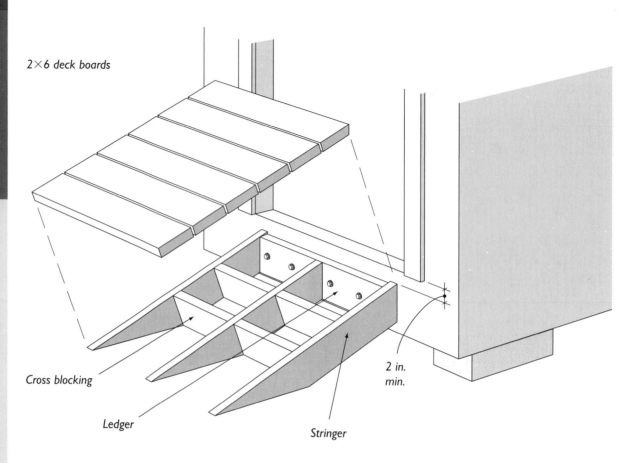

2×6 deck boards

Cross blocking

Ledger

Stringer

2 in. min.

Building a Ramp
Ramps usually have two or three PT 2×6 or 2×8 support stringers trimmed so they sit flush on the ground and fit against the shed. A 2× ledger supports the upper end of the frame, while cross blocking between the stringers helps support the decking. The ledger is fastened to the shed framing about 2 in. below the door's threshold; the decking is fastened to the stringers with 3-in. galvanized screws (which makes it easier to remove the decking if you need to unbolt the ramp from the shed some day).

What I like best about this type of step is that you can make the platforms virtually any size and in any configuration you like. For example, you can fan out the platforms, offset them by varying amounts, stagger each platform in opposite directions, extend them across the width of the building, or wrap the bottom platform around the corner of the shed. When you're done, you'll have a stylish multilevel entrance deck, not simply a set of stairs.

Ramps

If you've ever tried tugging a wheelbarrow up a set of stairs, you know why ramps are so valuable. A wooden ramp provides the only safe and sensible way to store lawn mowers, garden tillers, drop spreaders, garden tractors, and many other wheeled devices. Best of all, you can build a sturdy ramp in about an hour.

The length of the ramp will depend on the height of the shed floor; the higher it is, the

Arrange several large, relatively flat rocks at the doorway of a shed to create a natural-looking stepping-stone entrance. (Photo © Joseph Truini.)

longer you need to make the ramp. As a general rule, a 6-ft. ramp works well in most situations.

Stone steps

I think that natural stone is the best material for building shed steps. It's very attractive, naturally slip resistant, and extremely durable.

You can make a step from a single slab of stone or use several large rocks to create a stepping-stone entrance. These methods are especially well suited to situations where you need to step up only once to enter the shed.

However, large stones are extremely heavy and virtually impossible for the average person to transport and set in place by him or herself. The best approach is to buy the stone from a local quarry or masonry supplier and have the dealer deliver and install it.

A Moveable Ramp

One problem with a wooden ramp is that it doesn't create the most stylish entrance for an outbuilding. Most homeowners much prefer the look of steps or platforms, but they still occasionally need a ramp to wheel machinery into the shed. The solution is to make a removable ramp from a couple of 2×12s.

To create a smooth transition from the ramp to the shed floor, attach metal ramp fixtures to the 2×12s. The metal fixtures are sold at most home centers, hardware stores, and auto-parts

Make a pair of removable ramps with a set of extruded-aluminum ramp fixtures. Simply bolt each fixture onto the end of a 2×12. (Photo © Joseph Truini.)

dealers. Bolt one fixture to the upper end of each 2×12 with the hardware provided. Then whenever you need to wheel in a lawn mower or a wheelbarrow, just set the ramp in the doorway, with the lip of the metal fixtures resting on the shed's floor.

The length of a ramp is determined by how high the shed floor is from the ground. The higher the step, the longer the ramp must be. Use the chart below for guidance when building a ramp for your shed. Keep in mind, however, that longer ramps form gentler inclines that are safer and more comfortable to walk on. To determine how long your ramp should be (L), first determine the height from the doorway to the ground (H).

Ramp Height	Ramp Length
12 in.	4 to 4½ ft.
14 in.	4½ to 5 ft.
16 in.	5 to 5½ ft.
18 in.	6 to 6½ ft.
22 in.	7 to 7½ ft.
24 in.	7½ to 8 ft.

Building

CHAPTER THREE
Materials

I n between the time you design your shed and the time you build it, you'll need to specify all the building materials. The importance of this step is often overlooked, but each decision has an effect on the shed's style, longevity, and cost.

The previous chapter included information about supplies, such as the lumber, foundation materials, and wiring, that don't necessarily influence the look of a shed. This chapter explores some options for the more-visible elements, including siding, roofing, doors, and windows.

The building materials you choose will often be based on personal preference, but there are practical considerations as well. You may love the look of a cedar-shake roof, but if you live in a very humid climate or if the site is heavily shaded, wood may not be the best roofing choice; it tends to mold if it remains damp. Consider your choices carefully and you'll be rewarded with a long-lasting shed that's well within your budget.

PRO TIP

Planning to paint your shed? Save yourself some time and trouble by installing preprimed siding. It comes ready to paint with a factory-applied coat of primer.

TRADE SECRET

To ensure that a paint or stain finish won't peel or fade prematurely, coat the back of each piece of siding before you install it. This technique, known as back-priming, seals the boards and prevents moisture from passing through the siding from the back surface, which can blister the top coat finish. Also, be sure to apply paint or stain on all the exposed edges and the ends of each piece of siding.

(Photo © Western Red Cedar Lumber Association.)

IN DETAIL

Capture the rustic charm of a mountain cabin with wavy-edged bevel siding. This grade of western-red-cedar siding has an eye-catching undulating butt edge that creates interesting shadow lines across the building. Its roughsawn texture is peppered with tons of tight knots, making it ideal for finishing with stain or clear wood preservative.

Siding

The type of siding you choose will help define the shed's style as rustic or refined, casual or classic. Siding can be installed horizontally, vertically, or diagonally. Before you make a final decision, however, consider how you plan to finish the siding. Some types take paint really well; others should be stained or coated with clear preservative.

Cost and speed of installation are also valid concerns when choosing siding material. For example, cedar shingles are attractive but rather time-consuming to install. Plywood siding, on the other hand, is reasonably priced and goes up very quickly. Also, some siding types must be nailed to plywood wall sheathing. For other types, you can eliminate the plywood and nail the siding directly to the wall studs.

Wood siding, in all its variations, is by far the most common siding material for sheds and outbuildings. It's relatively affordable, long-lasting, and easy to install. Below are brief descriptions of the six most popular types of wood siding.

Bevel siding

Commonly called clapboards, bevel siding comes in long, thin planks that are installed horizontally on exterior walls. It's called bevel siding because the individual boards are cut with the sawblade tilted at a slight bevel angle to produce planks that are thinner at one edge than the other. As each course of siding is nailed up, the thin upper edge is overlapped by the thicker bottom (butt) edge of the course above it. This is the kind of siding installed on the Saltbox Potting Shed (see p. 106).

How Much Siding Do I Need?

Use the chart below to calculate how much bevel siding you'll need for each wall of your outbuilding. Note that information is given for five widths of siding and that the formula assumes a 1-in. overlap between the siding courses.

1. Calculate the total wall area by multiplying the width times the length.

2. Subtract the square footage of the windows and doors to determine the actual wall area.

3. Add 10 percent of the wall area for cutoff waste.

4. Multiply the sum by the appropriate linear feet factor listed below.

Siding Width	Linear Feet Factor
4 in.	4.80
6 in.	2.67
8 in.	1.85
10 in.	1.41
12 in.	1.14

Example:

1. An 8-ft. by 10-ft. wall has a total area of 80 sq. ft.

2. There are two 2-ft. by 3-ft. windows, for a total of 12 sq. ft. Subtract 12 from 80 to get 68 sq. ft.

3. Add 6.8 sq. ft. to 68 for a sum of 74.8 sq. ft.

4. Multiply 74.8 by 2.67 (for 6-in. siding). You'll need approximately 200 linear feet of siding to cover this one wall.

Most bevel siding is milled from western red cedar or redwood, two softwood species that are naturally resistant to rot and wood-boring insects. Spruce, cypress, and pine versions are also available in some regions. Bevel siding comes in several grades, from clear to knotty, and is typically smooth on one side and roughsawn on the other. If you plan to paint the siding, install the smooth side facing out. For a stain finish, install the rough side out; the stain will soak much deeper into the roughsawn surface and, therefore, last longer.

Bevel siding is available in several widths, ranging from 4 in. to 10 in. The most popular size by far is ½-in. by 6-in. siding, which actually measures about ⁷⁄₁₆ in. thick by 5½ in. wide. Courses are typically overlapped by at least 1 in., depending on how wide the boards are. For example, ½-in. by 6-in. siding should have no more than 4½ in. of wood exposed to the weather.

Use galvanized siding nails or, better yet, stainless steel nails to fasten bevel siding. Be sure to drive the nails through the plywood wall sheathing and into the wall studs. Otherwise, the nail tips will protrude into the shed's interior and create dozens of pointy, painful hazards.

Tongue-and-groove boards

Pattern siding is a broad family of wood boards that have been machined with various interlocking or overlapping joints. The most popular is V-jointed tongue-and-groove siding. This is the type of siding used on the Colonial-Style Shed (see p. 136).

Each ¾-in.-thick board is milled with a tongue along one edge and a groove along the other. When the boards are nailed up, the tongue of one board fits tightly into the groove of the adjacent board. Generally, an edge on at least one side of the board is also chamfered at a 45-degree angle, creating a decorative V-shaped joint along the seams when the boards are fitted together. This

Cedar bevel siding (commonly called clapboards) tapers in thickness from ⁷⁄₁₆ in. at the butt edge to about ⅛ in. along the upper edge. (Photo © Joseph Truini.)

There are many varieties of cedar bevel siding, including a rustic, roughsawn knotty grade that's ideal for finishing with stain. (Photo © Western Red Cedar Lumber Association.)

Tongue-and-groove, V-jointed siding can run horizontally; just be sure to install the boards with the tongue edges facing up. (Photo © Western Red Cedar Lumber Association.)

type of siding is typically installed vertically, but it can also be run diagonally or horizontally.

Like other types of pattern siding, tongue-and-groove boards are nailed directly to the wall framing; there's no need to first sheathe the walls with plywood. However, when installing it vertically, you must add horizontal blocking to the wall framing to provide solid nailing support for each

PRO TIP

For superior holding power—and to prevent nail pops—attach wood siding with stainless steel or hot-dipped galvanized nails that have spiral or ringed shanks.

TRADE SECRET

Tongue-and-groove siding can be installed horizontally, diagonally, or vertically. Just remember when installing it horizontally or diagonally, you must orient each board with its tongue edge facing up. That way, the siding will effectively drain away rainwater.

(Photo © Joseph Truini.)

WHAT CAN GO WRONG

Installing wood siding too close to the ground will eventually lead to decay. That's because the boards will wick moisture out of the soil, and wood that remains damp won't last very long. This problem is particularly troublesome if the shed is sided with untreated pine. To prevent rot, keep the siding above the grass line. That way, air will be circulate around the boards. To keep rot at bay, apply a coat of stain or paint to the ends of the boards.

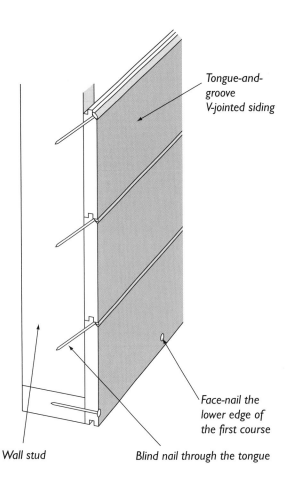

Tongue-and-groove V-jointed siding

Face-nail the lower edge of the first course

Wall stud

Blind nail through the tongue

Blind-Nailing Tongue-and-Groove Siding
Tongue-and-groove siding looks best when it's blind-nailed to the wall framing. Drive 2½-in. ring-shank siding nails at an angle through the tongue. The nail heads will be covered by the grooved edge of the next siding course.

board. Pattern siding can simply be face-nailed to the framing, but tongue-and-groove boards look best when the nails are driven through the tongue and then covered by the grooved edge of the next board, a technique known as blind-nailing.

Most tongue-and-groove siding is milled from western red cedar or redwood and has either a smooth or a roughsawn surface. It's readily available in 4-in.-, 6-in.-, and 8-in.-wide planks, with the 6-in. size being the most common. You can also buy tongue-and-groove boards made from untreated pine, but be aware that this product is very susceptible to rot—especially on the boards closest to the ground. If you choose untreated pine siding, protect it with stain or paint, then

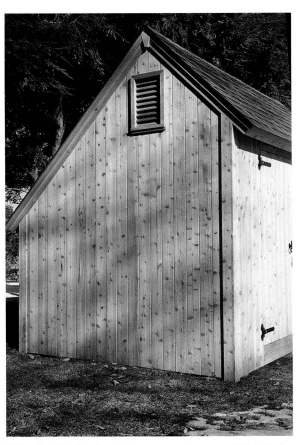

This shed is sided with tongue-and-groove, V-jointed cedar 1×6s. The knotty boards are finished with clear wood preservative. (Photo © Joseph Truini.)

reapply the finish at the first sign of water damage. You can also use tongue-and-groove pressure-treated pine planking as exterior siding. This product is designed for porch floors and isn't V-jointed, but it's highly resistant to rot and insects and costs much less than cedar and redwood.

Channel siding

This type of pattern siding is popular because of its versatility and rustic appearance. It's most often cut from knotty, roughsawn cedar and available in 6-in., 8-in., and 10-in. widths. Choose either 6-in. or 8-in. siding for small to medium-size sheds; select 10-in. boards for large outbuildings. Channel siding can be nailed up vertically or diagonally, but it is often installed horizontally to mimic traditional clapboards.

Channel siding can be installed vertically, diagonally, or horizontally. The roughsawn boards are available in widths of up to 10 in. (Photo © Western Red Cedar Lumber Association.)

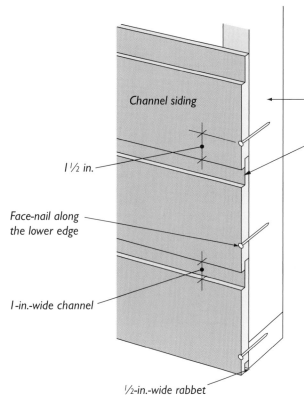

Channel siding

Wall stud

1 ½ in.

1 ½-in.-wide rabbet

Face-nail along the lower edge

1-in.-wide channel

½-in.-wide rabbet

Channel-Siding Installation
Fasten channel siding to the wall framing by face-nailing each course 1 ½ in. up from the bottom edge with 2-in. ring-shank siding nails (use one nail for siding up to 6 in. wide; two nails per stud for siding 8 in. and wider). The narrow rabbet on the bottom edge partially overlaps the larger rabbet on the top of the adjacent board, effectively holding the board in place.

The ¾-in.-thick boards have a modified rabbet joint cut into each edge. (A rabbet is simply an L-shaped notch running along the edge or end of a board.) A ½-in.-wide rabbet is milled from the rear surface along one edge, and a 1½-in.-wide rabbet is milled from the front surface on the opposite edge. When the boards are installed, the narrower rabbet on one board partially overlaps the larger rabbet on the adjacent board, resulting in a 1-in.-wide reveal, or channel, between the planks.

Board-and-batten siding

One of the oldest types of wood siding, board-and-batten siding consists of wide boards and narrow wood strips, called battens. The boards are nailed vertically to the wall framing, then the battens are installed to conceal the seams between the boards. Note that this type of siding can only be installed vertically.

Unlike other types of wood siding, there's no standard size for board-and-batten siding: You can make it from virtually any size lumber. However, the battens are usually cut from 1×3s and the vertical boards are cut from stock ranging from 1×8s to 1×12s. Again, narrower boards look best on smaller sheds; wider boards are used for larger buildings. For best results, cut the siding from a

rot- and insect-resistant wood, such as cedar, redwood, or pressure-treated lumber.

When you install the boards, leave about a 1-in. space between them so they can expand without buckling. If the boards are 6 in. or narrower, secure each one with a single row of nails

Capture the casual cottage look with board-and-batten siding, which consists of wide vertical boards and narrow wood battens. (Photo © Western Red Cedar Lumber Association.)

PRO TIP

To prevent siding from splitting as you nail it, first bore a pilot hole with a drill bit that's slightly smaller in diameter than the nail shaft.

TRADE SECRET

I've found that the quickest, most effective tool for staining siding is a 4-in. by 6-in. paint pad. The thin, flat pad allows you to apply an even, uniform coat without dripping stain all over the place. After spreading out the stain, let it soak in for about 10 minutes, then—and this is important—scrub the surface with a stiff-bristle scouring brush. This action accomplishes two things: It drives the stain deep into the wood grain and it removes any excess stain.

(Photo © Joseph Truini.)

TRADE SECRET

Add a bit of refinement to board-and-batten siding by routing decorative chamfers into the battens. Using a router fitted with a 45-degree chamfering bit, cut a ⅜-in.-wide chamfer into both edges of each batten.

driven through the middle of the board. If they're 8 in. or wider, attach them with two rows of nails.

After the boards are in place, fasten the battens over the gaps between the boards with a single row of nails. Make sure the battens overlap the boards by a minimum of ½ in. to provide adequate coverage in case the boards shrink.

A variation on this style of siding is called reverse board-and-batten siding. The battens are nailed up first, then the boards are installed to create a series of deep channels. Before you choose reverse board-and-batten siding for your shed, you should know that there's a much quicker and easier way to create a similar effect: Install channel siding vertically.

Cedar shingles

A good way to capture the look of a quaint country cottage is by siding your outbuilding with cedar shingles. The individual shingles are

These red-cedar shingles are finished with clear wood preservative to help retain the wood's natural beauty and auburn color. (Photo © Cedar Shake and Shingle Bureau®.)

installed in overlapping courses, which shed rainwater like water off a duck's back. The nails in each course are concealed by the shingles in the course above it, resulting in a very neat, natural appearance.

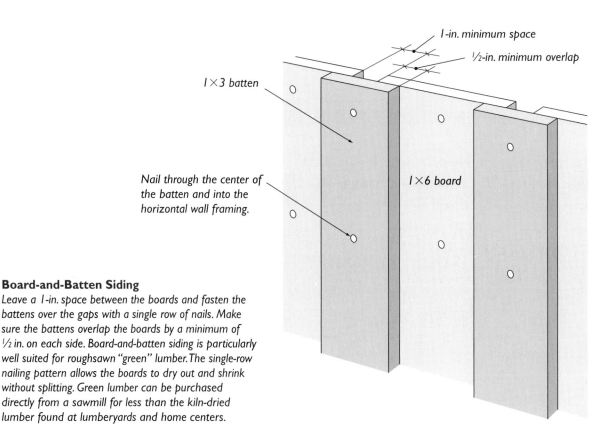

Board-and-Batten Siding
Leave a 1-in. space between the boards and fasten the battens over the gaps with a single row of nails. Make sure the battens overlap the boards by a minimum of ½ in. on each side. Board-and-batten siding is particularly well suited for roughsawn "green" lumber. The single-row nailing pattern allows the boards to dry out and shrink without splitting. Green lumber can be purchased directly from a sawmill for less than the kiln-dried lumber found at lumberyards and home centers.

1-in. minimum space
½-in. minimum overlap
1×3 batten
1×6 board
Nail through the center of the batten and into the horizontal wall framing.

Cedar shingles (bottom) are thinner and smoother than hand-split shakes (top), which are typically thicker and more roughly cut. (Photo © Cedar Shake and Shingle Bureau.)

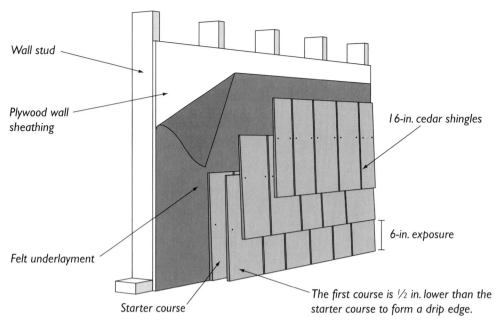

Wall stud

Plywood wall sheathing

16-in. cedar shingles

Felt underlayment

6-in. exposure

Starter course

The first course is ½ in. lower than the starter course to form a drip edge.

Cedar-Shingle Installation
Cedar shingles must be nailed to plywood wall sheathing. The nails in each row are concealed by the next course. A starter course of shingles is installed along the bottom of the wall to flare out the first row of shingles.

For most outbuildings, shingles aren't as popular as other types of wood siding. For one thing, cedar shingles are rather time-consuming to install and relatively expensive. Also, they can't be nailed directly to the wall framing; you must first install plywood sheathing. Finally, after the shingles are installed, there will be hundreds of nails poking through the walls and into the spaces between the wall studs. To keep from getting pricked accidentally, you'll have to cover the walls on the inside of the shed with plywood, wallboard, pegboard, or some other rigid material.

Like other types of wood siding, shingles can be finished with paint, clear wood preservative, or semitransparent or solid-color stain. It's not important which finish you apply, just that you apply one. Shingles last much longer and curl and split less when they're protected from the elements with a coat of finish.

Looking for a more rustic, less refined appearance? Consider installing hand-split cedar shakes. Shakes are similar to shingles, but they're much thicker and more heavily textured. Shakes can also be installed with a greater exposure to the weather, so you'll need fewer shakes than shingles to cover the shed.

+ SAFETY FIRST

Installing wood-shingle siding requires an awful lot of nailing. If you decide to speed the process by renting a pneumatic nailer, it's very important that you never allow anyone inside the shed while you're working. It's too easy for a nail to miss a stud and shoot off into the distance with dangerous results.

Shingle Styles

One reason you may want to consider using shingles is that they're available in a wide variety of custom-cut shapes, allowing you to add dramatic designs to an otherwise ordinary façade. Available from the same mills that sell standard shingles, these Fancy-Butt shingles comes in nine shapes, including diamond, round, half cove, hexagonal, arrow point, and fish scale.

Fancy-Butt shingles come in a wide variety of precut shapes for creating an amazing array of eye-catching sidewall designs. (Photo © Cedar Shake and Shingle Bureau.)

PRO TIP

Make sure that all vertical seams between sheets of plywood siding fall on the center of a wall stud. Fasten the sheets with 2-in. (6d) galvanized nails.

IN DETAIL

Redwood and cedar siding are often chosen for their natural beauty and handsome color. However, the only way to keep the wood from graying is to apply an appropriate color stain. For example, to maintain redwood's rich, crimson color, apply a coat of redwood-tone stain.

WHAT CAN GO WRONG

Tongue-and-groove siding is easy to install, but only if you protect the milled edges from damage. Use a tapping block cut from a scrap of siding to gently tap the boards into place.

TRADE SECRET

Save yourself hours of frustration by painting or staining plywood siding before you install it. It'll be much easier to apply the finish with the large sheets propped up on sawhorses.

Reverse board-and-batten plywood siding has a rough-textured surface machined with broad, widely spaced grooves. (Photo © Georgia Pacific.)

Grooved plywood siding is commonly available with the ⅜-in.-wide grooves spaced either 4 in. (shown) or 8 in. apart. (Photo © Georgia Pacific.)

Plywood siding

There are two valid reasons for siding your shed with plywood: it's economical and it covers a lot of area very quickly. Plywood siding is a ⅝-in.-thick exterior-grade product that's commonly available in 4-ft.-wide by 8-ft.-long sheets. It can also be special-ordered in 9-ft.- and 10-ft.-long panels.

Plywood siding comes in several styles, including roughsawn, primed, and unprimed wood, as well as something that resembles reverse board-and-batten siding. However, the most popular style by far is grooved plywood siding (though nearly everyone refers to it as T-1-11, the trade name of grooved siding manufactured by Georgia Pacific).

This type of siding has a roughsawn surface that features a series of equally spaced ¼-in.-deep by ⅜-in.-wide grooves. You can buy it with the grooves spaced either 4 in. or 8 in. apart. The style to choose depends on the size of your building. The 4-in. design looks best on smaller structures; larger outbuildings can accommodate the wider 8-in. pattern. I installed 4-in. grooved siding on the small Lean-to Shed Locker (see p. 76). The large Gambrel Storage Barn is sided with 8-in. grooved plywood (see p. 168).

Textured plywood siding comes in 4-ft. by 8-ft. sheets and provides the quickest, easiest way to cover large wall sections.

FIBER-CEMENT SIDING

Fiber cement is a unique siding material that looks like wood but is actually a mixture of cement, sand, and cellulose fibers. It has been around for nearly 12 years and is slowly gaining popularity with professional homebuilders. However, it's seldom used on outbuildings even though, in many ways, it's a near-perfect choice.

Fiber-cement siding won't rot, crack, delaminate, or burn. It's highly resistant to moisture, including salt air and salt spray. Wood-boring insects hate it, and it holds paint better than almost any other type of siding does. In most cases, it's even less expensive than quality wood siding.

However, fiber-cement siding isn't the world's most user-friendly building material. The siding is much heavier than wood and more difficult to nail because it's so hard. It's also nearly impossible to cut with ordinary tools. You can use a portable circular saw fitted with an appropriate carbide-tipped blade, but it'll spew out a cloud of dust thick enough to shut down an airport. You can also score and snap it with a carbide-tipped scoring tool (doesn't that sound like fun?), but the best way to cut fiber cement is with a pair of electric or pneumatic shears. The shears are very expensive to buy, but you should be able to rent a pair at a well-stocked tool rental outlet.

Fiber-cement siding comes in 5/16-in.-thick by 12-ft.-long planks and ranges in width from 5 in. to 12 in. The planks resemble wood clapboards; the material is also available in various shingle-style planks that go up quickly but look like individual shingles. Some manufacturers also offer styles that mimic grooved plywood siding; stucco; and specialty shingles with rounded, scalloped, and clipped-corner designs.

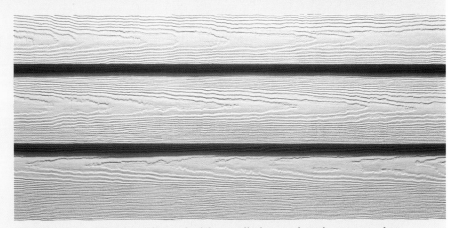

Fiber-cement siding is embossed with a realistic wood-grain texture that looks very much like roughsawn bevel siding. (Photo © James Hardie Building Products®.)

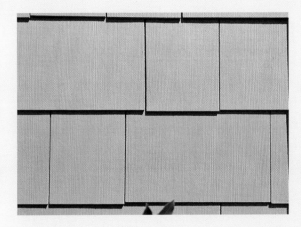

What appear to be individual wood shingles installed with staggered butt edges are actually 4-ft.-long fiber-cement planks. (Photo © James Hardie Building Products.)

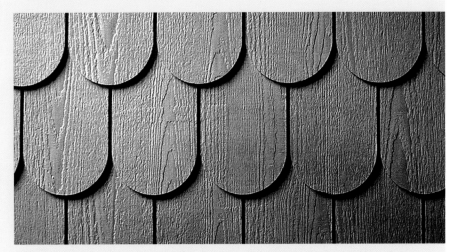

These attractive, rounded shingles are just one of the many varieties of specialty shapes now available in fiber cement. (Photo © Weatherboard.)

WHAT CAN GO WRONG

Leave a space of at least ⅛ in. between cedar shingles. These gaps, called keyways, allow the wood to expand as it absorbs moisture. If the shingles are butted tightly together, they won't have room to expand and will buckle and crack.

TRADE SECRET

To determine how much roofing you'll need, start by figuring out the area of the roof in square feet. However, when ordering the roofing, you'll have to ask for it by the "square." In roofing parlance, a square of roofing equals 100 sq. ft. Therefore, if your outbuilding has 250 sq. ft. of roof area, you'll need two and a half squares of shingles. It usually takes three bundles of three-tab shingles or four bundles of architectural-style shingles to equal one square.

Plywood siding is nailed directly to the wall framing. Shallow rabbets, similar to those on channel siding, are milled into the edges of the sheets. When the panels are installed, one sheet overlaps the next and creates a tight, water-resistant joint. To ensure that the plywood panels won't pull free over time, I like to apply a bead of construction adhesive along every stud before installing the plywood siding. That way, if the framing lumber shrinks slightly or if the nails work themselves loose, the siding will remain fast.

If you're a traditionalist or a woodworking purist, you'll probably be happier with another type of wood siding. Plywood siding, despite all its virtues, will always be a poor imitation of the real thing: solid wood siding.

Roofing

During the initial design phase of your outbuilding, you probably won't spend a lot of time fretting about which type of roofing to install. That's certainly understandable, as so much forethought must go into selecting the siding, windows, and paint color. However, don't underestimate the impact a roof can have on the style and appearance of an outbuilding.

There are a wide variety of roofing materials available, but most of them aren't very well suited for do-it-yourself installation. Here, I'll discuss the three most appropriate roofing options for the backyard builder, including two reliable

Asphalt roof shingles come in two basic types: three-tab shingles (top) and architectural-style shingles (bottom), which are also called laminated shingles. (Photo © Joseph Truini.)

standards—asphalt shingles and cedar shakes—and a modern version of an old classic: slate.

Asphalt shingles

To say that asphalt shingles are popular is an understatement of immeasurable proportions. I'd estimate that a vast majority of outbuildings—probably close to 95 percent—are covered with asphalt roofing. And why not? It's affordable, easy to install, readily available in a wide array of colors, and surprisingly durable; many brands carry up to a 40-year warranty.

There are two basic kinds of asphalt shingles: organic-based shingles and fiberglass-based ones. The organic type is usually less expensive and available from more manufacturers; fiberglass-reinforced shingles last longer, are lighter in weight, and are more fire resistant. You can find very good—and not-so-good—versions of both kinds at most lumberyards and home centers.

There are also two main categories of asphalt shingles: three-tab shingles and architectural-style shingles. Standard three-tab shingles represent a very basic, generic style. Each shingle is a single layer thick, with two narrow slots cut into it to create the three tabs.

Architectural-style shingles, which are also commonly called laminated shingles, consist of two strips of asphalt roofing, one laid on top of the other. This type of shingle has a solid bottom piece and a top strip notched with widely spaced, dovetail-shaped tabs. The two strips are laminated together at the factory. When installed, the laminated construction forms a heavily textured surface with deep shadow lines.

Architectural-style shingles are a dramatic improvement over three-tab shingles in terms of visual impact. The shingles are available in several muted colors that, in many cases, mimic traditional roofing materials. For example, the tan tones resemble weathered cedar shakes and the gray-black shingles look somewhat like slate.

Cedar roofing

Many roofing materials are more affordable, longer lasting, and easier to install than a cedar roof, but none can compare to its natural beauty and distinctive texture. The one disadvantage of any wood roof, of course, is that it offers very little resistance to fire, unless it's treated with a fire retardant.

There are two basic types of cedar roofing: shingles and shakes. Shingles are thinner, smoother, and more uniform. Shakes are thicker and rougher. Some shakes are sawn from logs;

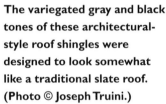

The variegated gray and black tones of these architectural-style roof shingles were designed to look somewhat like a traditional slate roof. (Photo © Joseph Truini.)

others are hand-split for a more rugged, uneven surface. (See p. 67.)

A shingle roof has a smart, elegant appearance, with clean lines and a low profile. The Saltbox Potting Shed (see p. 106) has a cedar-shingle roof. A shake roof is more rustic and robust-looking. Its surface has a lot of character and deep shadow lines.

A cedar roof is normally installed over spaced sheathing, not plywood. Spaced sheathing—also called open or skip sheathing—is simply a series

Shingles are uniformly sawn to about 3/8 in. thick, resulting in a stylish, low-profile roof that casts very thin shadow lines.

others are hand-split for a more rugged, uneven surface. (See p. 67.)

✔ According to Code

For fire safety reasons, some building departments have banned the use of wood roofs. The original building code was written to protect house roofs, but the code is typically applied to all structures, including outbuildings. If you're considering a wood-shingle or -shake roof for your shed, be sure to get approval from the local building inspector.

IN DETAIL

Here are five common roofing terms that every backyard builder needs to know:

- **Sheathing.** The plywood or solid wood layer to which the shingles are nailed.

- **Drip edge.** The metal flashing applied to the roof edges to deflect rainwater runoff.

- **Building paper.** An asphalt-saturated felt underlayment, sometimes called tar paper, that's stapled to the sheathing.

- **Starter course.** A single layer of shingles installed along the edges and ends of the roof prior to the first course.

- **Exposure.** The amount of shingle left exposed to the weather, usually about 5 in.

TRADE SECRET

If you're planning to install a wood roof, save yourself some trouble by ordering prefabricated ridge caps. The easy-to-install, V-shaped caps fit onto the ridge after the roof is covered with shingles or shakes. The caps are made from the same quality cedar, so they'll perfectly match the color and texture of the roof.

Hand-split cedar shakes have a rough texture and very thick, uneven butt edges that cast long, deep shadows across the roof. (Photo © **Cedar Shake and Shingle Bureau.**)

of 1×4 slats nailed across the rafters. The slats are spaced a few inches apart, which allows air to circulate behind the shingles or shakes to keep them dry. That's the key to a long-lasting wood roof; if the shingles can't breathe, they'll stay damp and quickly deteriorate.

Western-red-cedar shingles and shakes are widely available nationwide. In some areas you may also find eastern-white-cedar shingles, which are fine for sidewalls, but stick with western red cedar for roofing. It simply lasts longer and outperforms any other type of cedar.

Cedar shingles are available in several grades. For sidewalls and roofs, order Number 1 Blue

Roof Shingle Fungus Fighter

If a shed is built in a very shady location or in a region that receives a lot of rain, the roof will remain damp and dark. Under those conditions, it's only a matter of time before fungus starts to grow on the roof shingles. One way to help combat this problem is to trim back all nearby tree branches to permit sunlight to shine through. Unfortunately, that's not always a very practical or effective solution. A better answer is to install Shingle Shield™ protector strips along the roof near the ridge board.

These thin metal strips are made from 99 percent pure zinc, a proven fungus fighter. Each 3-ft.-long strip has an integral alignment guide and prepunched nail holes to simplify the installation. To install the metal strips, slide them under the roof shingles close to the roof peak, fastening them in place with a couple of short roofing nails. Every time it rains, water will run over the strips and release an invisible coating of zinc oxide. As the zinc oxide washes down over the roof, it'll be

absorbed by the shingles and inhibit the growth of fungus, moss, and algae. Shingle Shield can be installed on asphalt roofing and cedar shingles and shakes; according to the manufacturer, it will remain effective from more than 20 years.

Slip a zinc protector strip underneath an upper course of roof shingles to combat the growth of unsightly fungus and algae. (Photo © **Chicago Metallic®.**)

Label shingles. These premium-grade shingles are perfectly clear (no knots) and cut from all-heartwood for optimum decay resistance. They're available in 16-in., 18-in., and 24-in. lengths; the 16-in. size is suitable for most roofing jobs.

There are four premium grades of cedar shakes. My favorite choice for roofing is a grade designated as CERTI-SPLIT™ hand-split shakes. It has a rough-split face, but the smooth-sawn back makes installation easier. It's not likely that your local lumberyard will carry all grades of shingles and shakes, but it will be able to special-order any items that aren't in stock.

Note that some grades of shingles should be installed only on sidewalls. Never use those grades on a roof; they simply won't last as long. Before you specify a specific grade, check with the lumberyard or the Cedar Shake and Shingle Bureau (see Resources on p. 198) for more information.

Faux slate

Real slate roofs are durable and beautiful; but aren't typically used for shed roofs because they're difficult to install and expensive. Dura Slate Roofing System™ shingles are a new version of this ancient roofing material. This imposter resembles real slate—even up close—but it's made of a polymer compound similar to hard rubber.

The 12-in. by 18-in. shingles won't rot, dent, or split. They are flexible enough to withstand rough handling, are extremely water-resistant, and come with a 50-year warranty. The shingles are resilient but can easily be cut with a utility knife—just score and snap them. Faux-slate shingles are installed in a manner similar to that of asphalt shingles: They are fastened to plywood sheathing with standard roofing nails. Although faux slate isn't cheap—it is usually two to three times more expensive than asphalt shingles—the installation

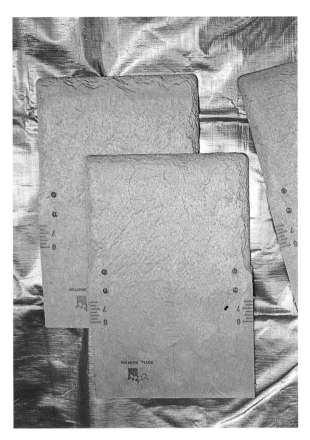

Looking very much like real slate, these 12-in. by 18-in. Dura Slate Roofing System shingles are made of tough polymer similar to hard rubber. (Photo © Joseph Truini.)

The realistic muted gray colors and embossed surface of this faux-slate roof are almost indistinguishable from real slate. (Photo © Joseph Truini.)

PRO TIP

Don't make the door until after the shed has been framed and the siding has been installed. Then custom-build the door to fit the opening precisely.

WHAT CAN GO WRONG

Don't wait until the day you nail down the plywood roof sheathing to go buy the shingles. Most lumberyards and home centers stock only three or four of the most popular shingle colors. If you want something a little out of the ordinary, such as forest-green shingles, it'll probably have to be special-ordered, which could take anywhere from a couple of days to a few weeks. I usually order the roofing on the same day that I pick up the framing lumber.

WHAT CAN GO WRONG

The basic reason why a roof doesn't leak during a rainstorm—even though it's made from hundreds of individual pieces—is that the shingles overlap each other. When installing a wood shingle or shake roof, be sure to stagger the vertical seams from one course to the next by a minimum of 1½ in. That way, when water runs between two shingles, it'll hit the shingle below and drain off the roof.

time, tools, and degree of difficulty are basically the same for both types of roofing.

I installed a Dura Slate roof on the Lean-to Shed Locker (see p. 76) and noticed that the shingles were much easier to cut and nail after they were warmed by the sun for a few minutes. When cold, the polymer shingles turn hard and are much more difficult to score with a utility knife.

The one telltale sign that Dura Slate is not the real thing is that the shingles are a little too perfect—a bit too uniform in size, shape, and color. Real slate is slightly more irregular, with scaly surfaces and corners of varying shapes. Still, Dura Slate makes a great-looking, long-lasting roof that any homeowner can install with ordinary hand tools. It's available in four real-slate colors: gray, black, green, and mulberry (dull red).

Doors

There are no hard and fast rules to follow when it comes to selecting a particular type of door for an outbuilding. As long as it opens smoothly and latches securely, it doesn't much matter what style the door is. Begin your search for the right door by taking a very practical approach. First, decide whether you need a single or a double door. A single door is adequate for most sheds, but if you're going to store a garden tractor, boat, or trailer, you'll need to install a pair of double doors.

Next, choose between a hinged swinging door and a sliding door. Swinging doors are easier to install and the hardware costs much less, but sliding doors glide completely out of the way and they won't accidentally slam shut. Another important consideration, especially if you live in a colder climate, is that hinged doors close more tightly and do a better job of blocking wind than sliding doors do.

Also, think about whether you want the door to discreetly blend in with the shed or to stand out as an architectural accent. To make the door

Batten doors can be used for single- or double-wide openings, such as this one, which has an overhead transom window. (Photo © Joseph Truini.)

inconspicuous, simply paint or stain it to match the siding. In fact, if you build it from the same material as the siding, it'll virtually disappear. If you'd prefer to create an eye-catching focal point, paint the door a bright color, use ornate wrought-iron hardware, or install elaborate architectural trim around the door opening.

Following are descriptions of four types of doors that you can readily build from commonly available lumber. There are no tricky joints to mill or complicated assembly sequences; each can be built with nothing more than a cordless drill/driver and a portable circular saw or table saw.

However, if you don't have the time or inclination to build a door from scratch, don't worry about it. Go out and buy one. Most home centers and lumberyards stock several styles of prehung doors made from wood, steel, and fiberglass. If you do take the buy-it-yourself approach, just make sure you select an exterior-grade door.

Batten doors

Carpenters have been building batten doors for more than five centuries. This style of door is still popular today—especially with do-it-yourselfers— because it looks great and is very easy to build.

A batten door consists of little more than a few vertical boards joined together with wooden strips, called battens. The battens are usually attached to the back of the door in an X- or a Z-shaped pattern, which reinforces the door and keeps it from sagging. You can make a door from standard, square-edged 1× boards, but it'll be much stronger and more attractive with tongue-and-groove, V-jointed cedar 1×6s. Plus, should the boards shrink slightly, no gaps will appear between them because of the interlocking tongue-and-groove joints.

It's pretty easy to build a batten door. First, start by cutting all the 1×6s to length. Then lay the boards face down on a flat surface and clamp them together. Next, saw off the tongue and groove from enough of the 1×6s to make the battens. Lay out the battens on the door, as shown in the drawing on p. 70, cut them to length, apply a bead of construction adhesive to each one, and screw them to the back of the door. For the average-size door, I recommend attaching three horizontal battens—one each at the top, center, and bottom—and two diagonal battens.

This handsome batten door is made from tongue-and-groove V-jointed boards, the same 1×6 boards used for the shed's siding. (Photo © Joseph Truini.)

The back view of this batten door reveals the positions of the three horizontal battens and the two diagonal Z-battens. (Photo © Joseph Truini.)

TRADE SECRET

Surface-mounted hinges are the easiest type of door hinges to install on an outbuilding, but they don't offer much in the way of security. Even with the doors padlocked, someone can get into the shed by simply removing the hinge screws. To thwart such thievery, replace at least one screw in each hinge leaf with a carriage bolt. If necessary, enlarge the hole in the hinge to accept the bolt, then add a washer and hex nut on the inside.

(Photo © Joseph Truini.)

WHAT CAN GO WRONG

Building shed doors often requires screwing through the back to attach a batten or face frame. Be very careful that the screws don't poke through the front. This can happen if you drive the screw head too far into the back surface.

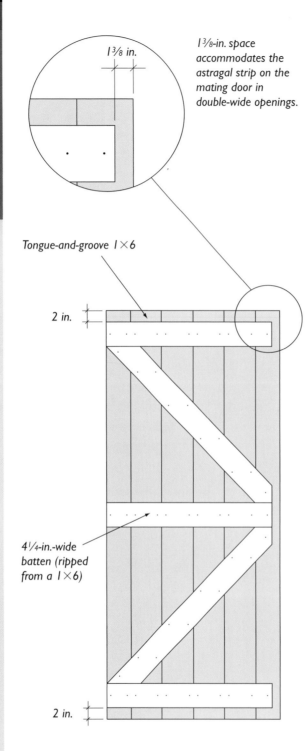

$1\frac{3}{8}$ in.

$1\frac{3}{8}$-in. space accommodates the astragal strip on the mating door in double-wide openings.

Tongue-and-groove 1×6

2 in.

$4\frac{1}{4}$-in.-wide batten (ripped from a 1×6)

2 in.

Batten Door
The rear view of a typical Z-batten door reveals three horizontal battens and two diagonal ones. The door panel is made from several tongue-and-groove boards.

A plywood door is made from a single plywood panel, which is then trimmed with a perimeter face frame cut from 1×4s. (Photo © Joseph Truini.)

Plywood doors

This style of door is the quickest and easiest type to build. It's simply a piece of plywood cut to fit the door opening, then framed with 1×4s to strengthen and stiffen the door. Not surprisingly, plywood doors are usually installed on sheds covered with plywood siding. And, of course, the same plywood used for the siding can be used for the doors.

These doors are often made with ⅝-in.-thick grooved plywood, but plain (not grooved) plywood siding with a roughsawn texture works just as well. For the frame, I like to use a weather-resistant wood, such as cedar or redwood.

Dutch doors

A Dutch door is easily recognized by its unique split personality: The top and bottom halves of the door swing independently of each other. The bottom section is often closed for a modicum of

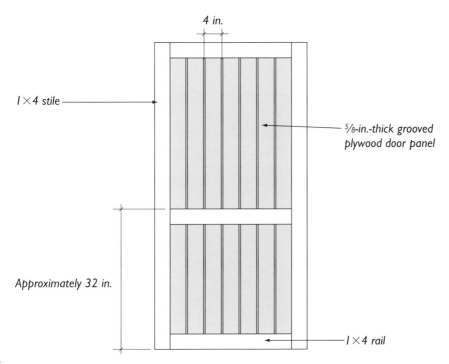

4 in.

1×4 stile

⅝-in.-thick grooved plywood door panel

Approximately 32 in.

1×4 rail

Plywood Door

One of the easiest of all shed doors to build, a plywood door is simply a grooved plywood panel with a 1×4 face frame screwed on from the back. Attach the face frame to the front of the plywood with construction adhesive and decking

screws. Drive screws from the back so the screw heads won't be visible on the front. Run the 1×4 frame around the perimeter of the plywood panel, then attach a horizontal 1×4 across the door at or slightly below the center point.

privacy and security, while the top half is left open for fresh air and neighborly chats. The two halves can also be locked together to operate as a standard swinging door. Although you probably wouldn't hang a Dutch door on a basic storage building, it'd be the perfect choice for adding a little old-world charm to a woodshop, backyard stable, or potting shed.

Dutch doors originated in (surprise!) the Netherlands during the early 1600s. They were first installed in front entryways before being relegated to secondary doorways at the side and rear of the house. Nowadays, Dutch doors are rare, though you may occasionally see one on a horse barn, garage, or roadside produce stand. The scarcity of these double-swingers isn't because people don't like them; in fact, most people love Dutch doors. You don't see them very often because Dutch doors are expensive to buy, few carpenters want to be bothered making them, and

Form and function come together in this charming Dutch door. Note the use of X-battens to create the cross-buck design. (Photo © Joseph Truini.)

TRADE SECRET

Here are three important rules to follow when making a plywood door:

- Use exterior-grade plywood that's a minimum of ⅝ in. thick.

- Attach the wood frame with weather-resistant galvanized or stainless steel screws.

- To reduce the likelihood that the plywood will bow or twist, apply two coats of stain or paint to both sides and to all four edges of the door.

IN DETAIL

An in-swinging door is easy to use because you can open it and step into the building with one motion—you don't have to step back away from the building, as you must with an out-swinging door. However, out-swinging doors are still preferred for storage sheds because they don't take up any interior space. The average in-swinging door requires about 15 sq. ft. of clear space in which to swing open.

even fewer homeowners know how to build one. I can't do anything about the first two reasons, but I can explain a couple of relatively simple ways to make your own Dutch door.

The first method is based on the design of a traditional batten door. Fasten together several tongue-and-groove boards with battens. Then saw the door in two and hang each half with a pair of hinges. For an out-swinging door, consider putting battens on both sides. That way, when the bottom half is closed and the top half is open, the two visible surfaces will match. (For more information on building a Dutch door, see p. 130.)

Here's an even easier way to build a Dutch door. Buy a solid-core, flush-panel exterior door and cut it in half. Protect the freshly cut ends by gluing on thin wood strips, then paint or stain the fronts, backs, and all the edges to seal out moisture. To create a little visual interest, rout shallow V-grooves in the door (spaced about 4 in. apart) or trim it with faux battens made from 3½-in.-wide strips of ¼-in. plywood.

Sliding doors

Long before the advent of sectional roll-up doors, nearly every carriage house, horse barn, and garage was outfitted with sliding wooden doors. Today, sliding doors still make sense for many larger outbuildings. They glide open with very little effort, can't be blown shut by the wind, and cover larger openings than do standard swinging doors. Because sliding doors fit over—not into—the doorway, you don't have to spend any time trimming and fussing to make the door fit precisely.

You may want to consider a sliding door if you live in an area that receives significant snowfall. Unlike an out-swinging door, a sliding door can be opened the morning after a blizzard without

The sliding batten door on this woodshop features a large window that sports a decorative grille made of ⅝-in.-dia. rebar. (Photo © Joseph Truini.)

Sliding doors are hung from rollers that glide on a track. The round rollers of this cannonball hanger fit into a tubular track.

your having to shovel the snow out of the way first. That's a big advantage, especially if you store your snow blower in the outbuilding.

A sliding door is basically just a slab of wood suspended from an overhead steel track. Virtually any style of door can be used with a sliding track, including plywood, solid-core, and batten ones. The door is attached to the track with a roller mechanism. Two roller assemblies (called trucks) are required for each door. Quite a few companies make sliding-door hardware, which come in only two basic types: trolley hanger and cannonball hanger. The trolley hanger has a wheel assembly (similar to a roller skate) that fits into a square or rectangular track. The cannon-ball hanger has a ball-shaped roller mechanism that fits into a round tubular steel track. Both types work well, but I've found that the cannon-ball hanger rolls a bit more smoothly and won't jump off the track.

Whichever type of sliding-door hardware you choose, make sure it can be easily adjusted after the door is hung. Check the installation instructions or call the manufacturer. It's important that the hardware have both vertical and lateral adjustments.

Windows

Natural light, fresh air, and views are just three reasons for installing windows in sheds. They also add character and style to an otherwise ordinary façade. There are dozens of window types and sizes from which to choose. Most are made of wood, aluminum, or vinyl. Clad windows are built from wood but have an exterior covering—or cladding—of aluminum or vinyl.

Let the Sun Shine in

When you really want to flood an interior space with natural light, and ordinary windows won't do, consider installing roof windows. These narrow units fit between rafters framed on 24-in. centers and extend down to the eave. The 20-in. by 70-in. windows shown here feature aluminum frames and insulated (double-pane) glass panels.

This technique is most effective when you gang together three or four windows in a row to create a partial greenhouse. However, be aware that you'll have to install roller shades or some other sun-blocking device to prevent the interior temperature from getting too hot, especially if the windows are placed on a south-facing slope. To minimize solar heat gain, install the windows on a north-facing slope. Note that roof windows should be installed only on roof slopes ranging from 4-in-12 to 12-in-12. The windows tend to leak on shallower-pitched roofs, which don't shed water as quickly.

Roof-mounted windows provide an effective way to transform part of an outbuilding into a year-round greenhouse. (Photo © Jerry Kolesar.)

PRO TIP

Salvage yards are great—and sometimes affordable—places to find interesting old windows, doors, hardware, and architectural millwork for outbuildings.

TRADE SECRET

When it comes to windows, are you tired of being square? Well then, shake things up a bit. Install a triangular, trapezoidal, or round-top unit. Or create an Eastern-inspired façade with a circular window.

IN DETAIL

Here are brief descriptions of six common types of windows:

- **Double hung.** A window with two sash, each of which slides up and down.
- **Single hung.** A two-sash unit, but only the bottom sash is operable.
- **Casement.** A crank-operated window with a side-hinged, swing-out sash.
- **Hopper.** A small, usually horizontal window that hinges at the bottom and tilts in at the top.
- **Awning.** A top-hinged window that swings out at the bottom.
- **Picture.** A large, flat, fixed (non-opening) window often flanked by narrower casement or double-hung units.

The type of outbuilding you construct will dictate the size, type, and number of windows you should install. For example, a single double-hung wooden window may suffice for a small storage shed, but a woodshop or artist's studio will be better served with several tall, light-grabbing casements.

Generally speaking, any window that you'd put in a house can be installed in an outbuilding. Here, I'll concentrate on the three types most commonly used for backyard buildings: wooden barn sash, aluminum sliders, and fixed transoms.

Wooden barn sash

As its name implies, a wooden barn sash is used in barns, stables, chicken coops, and other farm buildings. It's the most rudimentary and affordable of all window types. Technically, a barn sash isn't even a window; it's just a window sash comprised of glass panes set in a simple pine frame.

Before installing a barn sash, you'll need to build a window frame for it. I typically build frames out of 1× cedar, redwood, or pressure-treated lumber and add stops (narrow wood strips) to the inside of the frame to hold the sash in place. For ventilation, I like to allow the sash to tip in toward the shed interior.

Barn sash are commonly available at lumberyards and farm supply stores in 2-ft. by 2-ft. and 2-ft. by 3-ft. sizes. It's a good idea to apply a coat of paint or stain to all wooden surfaces of the sash, then let it dry thoroughly before setting it in the frame. Note that barn sash look best when they're painted a color that contrasts with the siding.

Aluminum sliders

Many architectural purists will scoff at the thought of putting a modern aluminum window on a traditional-style outbuilding. And they're right. These windows may look out of place on a colonial-style saltbox sided with cedar shingles.

A barn sash is a true divided-lite window, with single glazing set in an unfinished pine frame. Shown above is a 2-ft. by 2-ft. unit. (Photo © Joseph Truini.)

However, for many other outbuildings, aluminum sliders are ideal.

I like this style of window because the aluminum frame never needs painting, the product comes with insect screens, and, when it's open, the sash is contained within the frame—it doesn't swing out or lean into the shed. Aluminum sliders come in dozens of sizes, but the two most popular models for outbuildings are 2-ft. by 4-ft. and 2-ft. by 6-ft. units. Sliders are also easy to install: Just set the window in the rough opening from the outside, then have a helper on the inside drive a couple of screws through the aluminum jamb and into the wall framing on each side of the window.

Fixed transoms

This long, narrow window is installed over a doorway. It doesn't open for ventilation and emits a relatively small amount of daylight, but it's a valuable asset just the same.

A transom window adds a touch of elegance and visual interest to any entryway. It's one of the many small details that you can incorporate into

To add a little style and visual interest, this 2-ft. by 4-ft. barn sash window was painted dark green to contrast with the stained cedar siding. (Photo © Joseph Truini.)

Sliding aluminum windows are very durable, never need painting, and come in several sizes. Shown here is a 2-ft. by 6-ft. unit. (Photo © Joseph Truini.)

your shed design to make it unique. Transoms typically range from about 8 in. to 12 in. high and are as wide as the doorway. They look best when installed over a pair of double doors, but there's no reason you can't put one over a single door.

Transom windows are available from most major window manufacturers, but it's much cheaper to simply make one from scratch. Start by cutting a wooden frame from 1× stock; rip the pieces to about 2 in. wide. Rout a ³⁄₁₆-in.-deep by ⅞-in.-wide rabbet in the back of the frame pieces. Assemble the frame, then order a piece of double-strength glass to fit in the rabbets. Secure the pane to the frame with clear silicone adhesive. You can install the window as is, or cut short vertical wood strips to create the look of a divided-glass sash.

Transom windows add some light to a shed's interior but are usually installed as an architectural accent for a doorway. (Photo © Joseph Truini.)

Lean-to Shed

CHAPTER FOUR

Locker

1 Timber-Frame Foundation, p. 78

2 Wall Framing, p. 86

3 Roof Framing, p. 90

4 Siding and Trim, p. 92

5 Roofing, p. 96

6 Doors, p. 101

7 Interior Shelves, p. 105

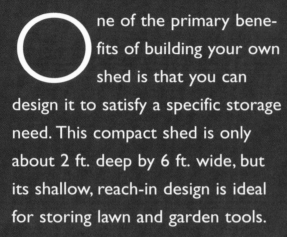

One of the primary benefits of building your own shed is that you can design it to satisfy a specific storage need. This compact shed is only about 2 ft. deep by 6 ft. wide, but its shallow, reach-in design is ideal for storing lawn and garden tools.

This style is known as a lean-to because it's built against an existing structure, such as a house, garage, or larger outbuilding. It features grooved plywood siding, double-wide doors and a beautiful faux-slate roof. The timber-frame foundation is inlaid with red-brick pavers.

Note that the timber frame extends 2 ft. beyond the shed to create a small brick "patio." This clever detail ensures that you'll always have a dry, level place to stand while retrieving a tool. The interior is outfitted with six wooden shelves, a door-mounted storage rack, and a large perforated hardboard panel for hanging long-handled tools. (To order a set of building plans for the Lean-to Shed Locker, see Resources on p. 198.)

TRADE SECRET

The wood preservatives used in pressure-treated lumber don't soak all the way into the middle of the board. Therefore, after cutting half-lap joints in 4×6 timbers, apply a generous coat of wood preservative to the freshly cut ends. Use a wide paintbrush and spread the preservative across the joint's cheek and shoulder cuts and on the end of the timber.

IN DETAIL

A half-lap joint is so named because it's cut halfway through the timber. For example, a 4×6 timber is actually 5½ in. thick. As a result, the half-lap must be cut 2¾ in. deep. The width of a half-lap joint always equals the full width of the timber. In this case, it's 3½ in., which is the actual width of a 4×6.

Timber-Frame Foundation

At the base of this 2-ft. by 6-ft. shed locker is a 4-ft.-wide by 6-ft.-long timber-frame foundation made of pressure-treated 4×6s. The four timbers are joined together with overlapping half-lap joints, then set on a bed of crushed granite. The stone base is necessary to prevent the timber frame from sinking into the soil. Once the frame is installed, 2⅜-in.-thick concrete-brick pavers are laid to create a hard-wearing, all-weather floor.

Assemble the frame

I used pressure-treated 4×6s rated for ground contact to build this timber-frame foundation, but 4×4s or 6×6s would have worked just as well.

Because the ground was relatively flat, I needed only one course of timbers. However, if your building site slopes more than 5 in. or so, you'll have to stack two or more timbers at the lowest end. If the ground slopes only 3 in. or 4 in. across the width of the foundation, you should be able to level the frame by partially burying the high end in the soil.

The first step is to cut the timbers to length to create the proper size frame. However, the exact frame dimensions are based on the size of the brick pavers used for the shed floor. That's because you want to avoid having to cut any bricks. It's much easier to adjust the size of the frame to accept an exact number of whole bricks. The frame doesn't need to be precisely 4 ft. wide by 6 ft. long; it can be a few inches larger or smaller. Its exact size isn't critical.

Sizing a Timber-Frame Foundation

One of the many indisputable facts of nature is that wood is easier to cut than concrete. Therefore, it's important that you size the timber-frame foundation to accommodate the concrete-brick pavers you'll be using for the shed floor. Here's how to determine exactly what size to build the frame so you won't have to cut any of the pavers.

Start by laying out a long, straight row of the brick pavers on a flat surface, such as a driveway or garage floor. Press the bricks tightly together to form a row that's approximately 72 in. long. Lay a second row of bricks perpendicular to the first one until it measures about 48 in. (as shown). Remember, the floor frame doesn't have to be exactly 4 ft. wide by 6 ft. long, but it must be able to accept a whole number of bricks.

Once you have the proper number of bricks laid out, measure each row and add ¼ in. for clearance; these measurements represent the inside dimension of the frame. To obtain the overall size of the frame, add 7 in. to each dimension (this is to accommodate two 4×6s, which each measure 3½ in. wide). Double-check your measurements and math, then cut to length the four 4×6 timbers. Note that this layout technique will work for concrete pavers of any size or shape, including octagonal, square, and rectangular ones.

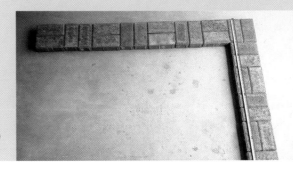

(Photo © Joseph Truini.)

Lean-to Shed Locker

This drawing reveals the construction method used to build the Lean-to Shed Locker. Note the timber-frame foundation, brick-paver floor, and plywood siding.

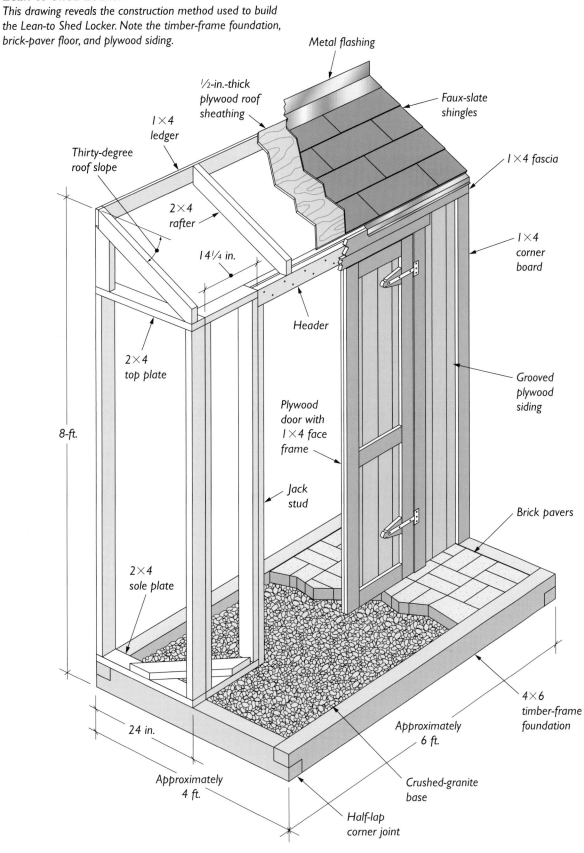

Metal flashing

½-in.-thick plywood roof sheathing

Faux-slate shingles

1×4 ledger

Thirty-degree roof slope

1×4 fascia

2×4 rafter

14¼ in.

1×4 corner board

Header

2×4 top plate

Grooved plywood siding

8-ft.

Plywood door with 1×4 face frame

Jack stud

Brick pavers

2×4 sole plate

24 in.

Approximately 4 ft.

Approximately 6 ft.

Crushed-granite base

4×6 timber-frame foundation

Half-lap corner joint

PRO TIP

Store bags of crushed granite in a dry location until you're ready to use them. If the bags get wet, they'll become extremely heavy and difficult to lift.

TRADE SECRET

After compacting the stone base, check it for level. Add stone to any low spots and scrape off a little stone from the high areas until all four sides of the base are perfectly level. Taking the time now to level the stone base will make it much easier to level the timber-frame foundation. If you do add stone to a low area, be sure to compact it with a hand tamper to prevent it from settling.

IN DETAIL

This particular shed locker, which is about 2 ft. deep by 6 ft. wide, was specifically built to store lawn and garden hand tools. However, the shed's basic design and dimensions can easily be altered to store a lawn mower, bicycles, carpentry tools, sporting goods, or swimming pool supplies.

Clamp together all four 4×6 timbers. Make the initial shoulder cut for the half-lap joints 3½ in. from the ends of the timbers.

Make a series of cuts, spaced about ¼ in. apart, through the waste-wood area; be careful not to saw beyond the shoulder cut.

For example, the bricks I used to create the floor for this shed locker measure 3⅞ in. wide by 7⅞ in. long. To accept a whole number of bricks, I cut the frame to 46¾ in. wide by 78⅜ in. long. If you're using a different size brick, you'll have to calculate the dimensions of the frame before cutting the 4×6 timbers.

I used a portable circular saw to cut the four timbers to the proper length. A 7¼-in. circular saw has a maximum cutting capacity of about 2¼ in.; a larger, 8¼-in. model will cut approximately ⅝ in. deeper. To cut clean through the 3½-in.-thick

+ SAFETY FIRST

Be sure to wear safety glasses, gloves, and a dust mask when sawing pressure-treated wood. Toxic chemicals are used to treat the lumber, and you don't want to inhale any of the dust. In fact, you should wear safety gear when sawing any type of lumber, because wood dust and chips of any kind can irritate your lungs and injure your eyes.

timber, you'll need to make one cut, flip over the timber, then make a second pass from the other side.

After trimming the four frame pieces to length, set them on sawhorses and prepare to cut a 2¾-in.-deep by 3½-in.-wide half-lap joint in each end of the four 4×6s. Instead of cutting each timber individually, clamp them together with their ends flush, measure 3½ in. from the end, and use a framing square to mark a square line across them. After adjusting the circular saw to its maximum depth of cut, make a shoulder cut along the pencil line. Then make a series of closely spaced cuts through the waste-wood area, being careful not to saw beyond the initial shoulder cut.

Remember that the circular saw will cut only about 2¼ in. deep, but the half-lap joint needs to be 2¾ in. deep. Therefore, use a handsaw to extend each circular-saw kerf to the bottom of the joint. Once that's done, use a hammer to knock away all the waste-wood slices, then use a wide-blade wood chisel to scrape the bottoms of the joints smooth.

Now remove the clamps and assemble the four timbers to form a rectangular frame. After treating the cut ends with preservative, overlap the 4×6s at the corners, making sure the half-lap joints fit together tightly, and check the frame for square by measuring the diagonals. When the two dimensions are identical, the frame is square. Fasten the 4×6s together by driving two 4-in.-long landscaping screws through each half-lap corner joint. You could join the timbers with galvanized lag screws or landscaping spikes, but the hex-head landscaping screws are much quicker and easier to install.

After checking to make sure the timber frame is still square, screw a 1×4 diagonally across it. This temporary brace prevents the frame from getting knocked out of square when it's later moved into place.

Smooth the bottom of the joints with a wide, sharp wood chisel. Hold the chisel flat with its beveled edge facing up to help prevent the tool from digging into the timber.

Prepare the site

It doesn't take long to prepare the site for a shed foundation that measures only 4 ft. by 6 ft., but that doesn't mean you can rush through the steps. To ensure that the timber frame provides long-lasting support for the shed and brick floor, you must thoroughly complete each step before moving on to the next one.

Start by marking the outline of the timber frame on the ground where you plan to construct the shed. There are two simple ways to do this. You can measure the frame, then use a hoe or shovel to scratch the outline into the ground, or set the frame in position and sprinkle white flour around the perimeter to mark the outline.

Once the frame's outline is clearly marked on the ground, use a flat shovel to carefully remove grass from the area. Lay the strips of sod off to the side in a shady spot and water them periodically; you may need them later to fill in bare spots after the timber frame is installed. Next, use a hoe to excavate a 4-in.-deep by 8-in.-wide trench

Assemble the rectangular timber frame by overlapping the half-lap corner joints. Check to make sure the joints fit together tightly.

Verify that the frame is square by measuring the two diagonals to see whether they are equal, then join the corners together with 4-in.-long landscaping screws.

PRO TIP

Invest in a high-quality 4-ft. level. Bargain-bin levels give inaccurate readings, making it impossible to build level foundations and plumb walls.

TRADE SECRET

After using a screed to scrape the stone base flat and even, you'll end up with a pile of stone along the front edge of the timber-frame foundation. The easiest way to remove this excess stone without disturbing the compacted stone base is with a rectangular steel trowel. Move the screed out of the way and set the trowel on a flat, level area of the stone base. Slide the trowel under the excess stone, then lift it up and dump the stone into a bucket.

WHAT CAN GO WRONG

After you've excavated the site and compacted the stone base, immediately install the timber frame and lay the brick pavers. If you leave the stone base exposed, there's a chance that it'll be disturbed by rain, playful children, or curious pets.

around all four sides of the outline. Pile the soil from the trench in the middle of the excavated area. Smooth out the bottom of the trench, then check it with a 4-ft. level. It doesn't have to be perfectly level, but it should be relatively close, say, within 1 in. or less across the length of the trench. Add soil to any low areas and scrape away the high spots to level all four sides of the trench.

The next step is to create a solid base out of crushed stone. You can use several types of aggregate, including processed stone or plain gravel, but my favorite material for this job is crushed granite. It compacts tighter and forms a much stronger, more stable base than other aggregates do. Crushed granite is sold at home centers and garden shops and often comes in two grades: coarse and fine. The coarse variety is ideal for this

application, but if you can find only the fine-grade type, it'll work nearly as well. However, don't use sand for the base. It's not stable enough and will eventually settle or wash out from under the timber frame.

Pour about a 2-in.-thick layer of crushed granite into the trench, smooth out the stone with a hoe, then spray it with a garden hose. I find that the water helps the crushed stone compact more readily and mixes with the stone dust to create a cementlike mortar. Use a hand tamper to pound the stone flat. You can buy a cast-iron tamper at a hardware store, but a homemade version works just as well. To make one, take a 36-in.- to 42-in.-long 4×4 and screw a 10-in.-long 2×6 block onto one end. Center the block on the 4×4 and attach it with four 3-in.-long decking screws.

Line the trench with 2 in. of crushed granite or other processed aggregate that contains stone dust. Smooth out the stone with a hoe.

Spray the stone base with a garden hose, then pound it flat with a hand tamper made from a 4×4 post and a short 2×6 block.

Set the assembled timber frame on the stone base. Note that a diagonal 1×4 brace is used to hold the frame square.

Use a 4-ft. level to confirm that the frame is level. Remove stone from underneath any high spots and add stone to any low areas.

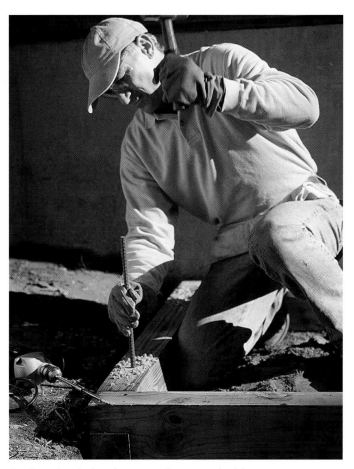

Anchor the timber frame to the ground with 18-in.-long pieces of ½-in.-dia. rebar. Drive in the steel bars with a small sledgehammer.

Install the timber-frame foundation

The next step is to carefully set the frame on top of the stone base. To determine its final resting spot, check whether it's aligned with the siding on the house. I do this by holding a 4-ft. level (or long, straight-edged board) against the siding that reaches the top of the timber frame. Then I move the frame in or out until its back edge aligns with the level, which typically places the frame about 1 in. to 1½ in. away from the foundation wall. Later, when the shed's wall framing is attached to the house, it'll land flush with the back edge of the frame.

Next, use a 4-ft. level to check all four sides of the frame, adding or removing some stone from underneath to make it level. It's a good idea to press down on each corner to make sure the

+ SAFETY FIRST

Get someone to give you a hand carrying the 4-ft. by 6-ft. timber frame to the building site. Although it's not all that heavy, it's a little cumbersome. You wouldn't want to drop it, twist it out of shape, or throw your back out while moving it.

PRO TIP

A timber-frame foundation will accept a variety of flooring materials, including marble chips, gravel, and concrete, which can be colored with masonry dye.

TRADE SECRET

Concrete pavers are extremely durable, but they're also porous, so they stain easily. After setting the shed floor, use a paint roller to apply a liberal coat of clear masonry sealer to the brick pavers. This protective finish will seal the pavers, making them much less porous and more resistant to stains.

WHAT CAN GO WRONG

If you live in an area that receives below-freezing temperatures, a brick-paver floor may be subjected to occasional frost heave. In extreme cases, the bricks can be pushed up enough to block the doors from swinging open. If that occurs, trim an extra ½ in. off the bottoms of the doors. By the way, once the ground thaws, the bricks should settle back down. If they don't, tap them down with a rubber mallet.

Circular Saw Safety Tips

A portable circular saw is an indispensable shed-building tool, but like all power tools, it's potentially dangerous. Heed the following precautions when using this popular power saw:

- Never operate a saw that has a frayed or cracked power cord.
- Check to make sure the lower blade guard glides up smoothly and retracts automatically.
- Immediately replace a dull or damaged blade.
- Always wear eye and ear protection.
- Use moderate pressure to slowly guide the saw through the cut. Never force the tool; let the blade do the cutting.

- Clean the board of all dirt, grease, and other debris before cutting.
- Before squeezing the trigger switch, make sure the sawblade isn't touching the edge of the board.
- Keep both hands on the tool while sawing; never hold the board across your knee and cut with one hand.
- Unplug the saw before adjusting the depth of cut or changing the blade.
- Never operate any power tool when you are fatigued or under the influence of alcohol or medication.

frame isn't rocking on any high spots. If you see any gaps below the frame, fill them with crushed granite. When you're happy with the frame's position, anchor it to the ground with long pins cut from ½-in.-dia. metal reinforcing bars (rebar).

Using a spade bit fitted onto an extension shaft, bore ½-in.-dia. holes for the rebar near each corner of the frame. Use a hacksaw or sabresaw equipped with a metal-cutting blade to cut four 18-in.-long pieces of ½-in.-dia. (No. 4) rebar. Then drive the rebar through the hole and into the ground with a small sledgehammer or an engineer's hammer. Tap the pins flush with the top of the 4×6, but be careful not to smash the frame with the hammer; otherwise, you may knock it out of level. After installing all four rebar pins, you can remove the 1×4 diagonal brace screwed across the timber frame.

Lay the brick floor

The last stage of building the timber-frame foundation is laying the brick floor. I like to use concrete pavers for floors rather than traditional clay bricks; concrete pavers are much harder and less

likely to crack. Most home centers carry three or four styles of concrete pavers, though they're made in a wide variety of shapes, sizes, and colors. For a better selection, visit a masonry supplier. For this project, I used a combination of dark red and charcoal gray pavers, each one measuring 2⅜ in.

Pound each brick paver into place with a rubber mallet. Set them in alternating right-angle pairs to create a basket-weave pattern.

thick by 3⅞ in. wide by 7⅞ in. long, and installed them in a basket-weave pattern.

The first step is to fill with crushed granite the area inside the frame to within 2 in. of the top. However, if you have 3 in. or 4 in. to fill, you can first add a couple inches of soil so you won't need as much granite. Be sure to pick out any rocks, roots, or other debris, then rake the soil smooth before compacting it with a hand tamper. Be careful not to hit the timber frame with the tamper.

Next, dump a few bags of crushed granite onto the compacted soil and smooth it with a garden rake, making sure you spread it into all four corners. Then thoroughly compact the stone. Spend at least three minutes pounding the surface flat, add a little more stone, then compact the surface again. Repeat this process until the stone is within 2 in. of the top of the frame, then add a final 1 in. or so of loose stone.

I use a screed to level the stone. To make one, cut a 2×3 or 2×4 about 3 in. longer than the length of the timber frame, then cut a 1×4 about 2 in. shorter than the inside dimension of the frame. Screw the 1×4 to the 2× board so that it hangs 2 in. below the 2×. To use the screed, set it on top of the timber frame and pull it toward you so the 1×4 scrapes away the excess stone and leaves a flat, level surface. If the screed skips over a low spot, fill the depression with stone, then smooth it with the screed again. Compact the surface one more time, add a little more crushed granite, then strike it off with the screed. Finally, use a garden hose to spray a fine mist of water across the stone. The surface is now ready for the brick pavers.

To set the bricks, start in one corner and work out in both directions. To create the basket-weave pattern, set the bricks in alternating right-angle pairs: Set two pavers parallel to the house, followed by two pavers perpendicular to it. Use a rubber mallet to pound each paver flush with the

Cover the compacted soil with an even layer of crushed granite. Then use a garden rake or hoe to smooth out the stones.

To create a solid, stable base for the brick floor, it's important that you thoroughly compact the stone base with a hand tamper.

Pull the screed across the timber frame. The 1×4 will smooth out the stone base and form a consistently uniform 2-in.-deep recess.

top of the frame. (See the photo on the facing page.) If a paver chips or cracks in half, replace it. Continue setting the brick pavers in this manner until the frame is filled. If the 4×6 timbers were sized and cut correctly, the last row of bricks will fit easily within the frame. If they don't fit, you'll have to cut them to size with either a motorized wet saw (available at tool rental shops) or a portable circular saw fitted with a masonry blade.

TRADE SECRET

When attaching the shed's end walls to the house, make sure they are plumb in two directions: left to right and front to back. To correct the front-to-back alignment, slip thin wood shims between the wall frame and the house siding. Place the shims high on the wall to tip the top forward. Put the shims down low on the wall to push out the bottom. Check the wall for plumb by holding a level on the stud at the front of the wall, not on the one against the house.

Sweep play sand across the surface to fill in the joints between the brick pavers and the timber frame and pavers.

Once all the pavers are set in place, pour a thick layer of fine-grain play sand onto the surface. Use a broom to sweep back and forth across the surface several times to drive the sand down between the pavers. The sand will fill any voids and help lock the pavers together. The excess sand should be swept up and saved. In about three or four months, after the original sand has settled or been washed out by rain, use the extra to fill in the joints again.

✓ According to Code

There's a building code that requires certain outbuildings to have a floor made of a nonabsorbent material. In other words, it can't be built from wood. Poured concrete and, in some cases, gravel are acceptable. Fortunately, this code refers only to large buildings that can be used as garages. It doesn't usually apply to tool-storage sheds.

Wall Framing

The walls and roof of this shed are framed entirely with pressure-treated 2×4s. You can use standard untreated 2×4s for much of the framing, but treated lumber offers greater resistance to water damage and wood-munching insects. However, you absolutely must use pressure-treated lumber for the sole (bottom) wall plates, the jack studs, and the rear corner studs that are attached to the house. Those parts are most susceptible to moisture damage.

Framing is usually nailed together; in this case, however, I used galvanized decking screws. The screws pull the parts together, hold tighter, and are easy to install with a cordless drill/driver. They're also less likely to split the lumber. Plus, you can always back out a few screws if you need to reposition a wall or move a stud.

Build the end walls

Start by building the shed's two 24-in.-wide end walls. For each wall, cut two 76¾-in.-tall wall studs and two 24-in.-long plates, one of each for the top and the bottom of the wall. I like to work

Fasten the 2×4 sole plate to the bottom of the end-wall studs using a cordless drill/driver and 3-in. decking screws.

Set the end wall on top of the timber-frame foundation and fasten it by screwing through the sole plate and into the frame.

Hold a 4-ft. level against the rear wall stud to ensure that it's perfectly plumb, then screw the wall frame to the house wall.

at waist level, so I lay the studs across sawhorses before attaching the top plates and sole plates with 3-in. decking screws. Note that it is much easier to drive screws when you drill pilot holes first.

Next, stand one end wall on top of the timber-frame foundation and press its rear stud flat against the house. Align the outer edge of its sole plate flush with the outside of the 4×6 frame. Then fasten the wall in place by driving three 3-in.-long decking screws through the sole plate and into the frame. Place one screw near each end of the sole plate and one in the middle.

Using a 4-ft. level, make sure the 2×4 frame is plumb—that is, perfectly vertical—in two directions: from side to side and from front to back. Then secure the frame to the house wall with four 3-in. screws. Place one screw near the top and bottom of the rear stud and equally space the other two screws in between. Repeat this process to install the second end wall.

Build the front walls

Once the two end walls are in place, it's time to construct the two narrow front walls, which are only 14¼ in. wide. Follow the same steps as above: For each wall, cut two 76¾-in.-tall wall studs and two 14¼-in.-long plates, then place the studs on sawhorses and screw on the top plates and sole plates. Note that the width of the two front walls affects the size of the doorway opening. In this case, the 14¾-in.-walls create a 40-in.-wide doorway. If you need a wider opening, make the front walls narrower; they can be reduced to a minimum of 6 in. wide.

Next, stand a front wall up against—and perpendicular to—one of the end walls you installed earlier. Although it's not necessary, I like to clamp the two walls together while I make sure the outside edges are flush and the tops of the walls are even (it may be necessary to slip shims under the front wall). When the walls are aligned, screw them together with four 3-in. screws, placing one screw near the top and bottom of the

TRADE SECRET

Use a power miter saw to cut framing lumber to size. It's quicker, safer, and much more accurate than a portable circular saw is. Plus, you can stack two 2×4s or four 1×4s and cut them simultaneously. This gang-sawing technique is especially useful when you're framing walls or crosscutting exterior trim boards. A 10-in. power miter saw can crosscut pieces of up to 5½ in.; an 8½-in. sliding compound-miter saw has a crosscut capacity of approximately 12 in.

WHAT CAN GO WRONG

Pressure-treated lumber is shipped fresh from the treatment plant in steel-strapped bundles. Weeks later, when the bundles are cut open at the lumberyard, the boards are often still soaking wet. If you use wet lumber, the building may twist, warp, and tilt out of square as the lumber dries and shrinks. Avoid these problems by hand-picking through the lumber pile to find the driest, straightest boards. You can tell the sodden ones simply by picking them up—they're much heavier than the dry boards.

Build the narrow front walls from 2×4s. Lay the studs across sawhorses and attach the top plates and the sole plates with 3-in. decking screws.

After confirming that the tops of the front walls and the end walls are absolutely even, fasten them together with 3-in. screws.

front wall stud and equally spacing the other two screws in between.

Unlike with the end walls (which sit on the timber frame), you can't drive screws through the sole plate of the front wall because it sits on top of concrete-brick pavers. Therefore, you must install

a diagonal 2×4 brace to anchor the front wall. Use a framing square to make sure the two walls form a 90-degree angle, then screw a 23-in.-long 2×4 brace diagonally across the two sole plates. The brace will add rigidity to the front wall and eliminate any wobbling that may occur when you

Add rock-solid stability to shed walls by screwing a diagonal 2×4 brace across the sole plates of the end walls and the front walls.

open or close the shed doors. Use this same construction sequence to install the second front wall.

Frame out the Doorway

The next step is to finish framing the doorway opening by installing the jack studs and the header. Cut two 2×4 jacks (sometimes called trimmer studs) to 76¼ in., and install one jack stud on the end of each front wall. Check each jack for plumb, slipping shims, as necessary, between the jack stud and the wall stud to keep it straight and plumb, then fasten the jacks with 2½-in. decking screws. Note that the jacks are cut 3½ in. shorter than the front walls to accommodate the header.

To make the header, first measure the distance between the two front walls. Then cut two 2×4s to that dimension and screw together the two boards to create the header. Set the header on edge on top of the jack studs. Check the header with a 4-ft. level and, if necessary, shim up one end to make it perfectly level. Then screw each end of the header to the front walls, driving two 3-in. decking screws through the front wall studs and into each end of the header.

On this shed, I installed a perforated hardboard panel (commonly known as pegboard) on the wall of the house over the siding. If you decide to install one of these handy tool-storage panels, you'll need to install the 48-in.-wide by 73-in.-long sheet now, because it won't fit through the doorway once the plywood siding is installed. First, though, mount two horizontal 2×4s along the back wall. Space them 48 in. apart and screw them to the end walls. Then slide the ¼-in.-thick perforated panel into position and fasten it to the 2×4s with 1¼-in. screws.

Note that perforated hardboard is also available in ⅛-in.-thick panels. However, for this application, use only the ¼-in. variety. The thinner panel is too flimsy to span such a large space.

Install a jack stud on the end of the two front walls. Note that it's 3½ in. shorter than the wall to accommodate the double 2×4 header.

Set the header on top of the two jack studs, then check it for level. If necessary, slip a thin shim under one end of the header.

Before you install the siding, slide the pegboard in place and screw it to the horizontal 2×4s attached to the rear wall.

IN DETAIL

The lower end of each rafter tail lands flush with the top wall plate. The simplest way to secure the rafter tail is with a single 2½-in.-long decking screw. First, bore a screw pilot hole through the rafter tail. Then drive the screw at an angle down through the rafter and into the top of the wall.

WHAT CAN GO WRONG

If you mistakenly cut a rafter a little too short, don't toss it onto the scrap heap. You may be able to save it. Cut a thin slice of wood equal in thickness to the amount you undercut the rafter. In other words, if the rafter is ¼ in. short, cut a ¼-in.-thick shim. Glue the shim to the ledger board, hold the undersize rafter against the shim, and screw it in place.

Roof Framing

The shed-style roof on this structure has a slope of 30 degrees, which is the perfect angle for this particular size building. A shallower slope wouldn't shed rain or snow as well and a steeper slope would significantly reduce the height of the doorway opening.

The roof frame consists of four 2×4 rafters and a 1×4 ledger board. To simplify the installation process, the rafters are first fastened to the ledger, then the entire assembly is lifted onto the shed walls. This prefabricated building method is not only faster and more accurate than the piecemeal approach, but it is also safer because you spend much less time working on a ladder. Note that the plywood roof sheathing and faux-slate shingles aren't installed until after the siding is completed.

Cut the roof rafters

The first step in building the roof is to cut one rafter to length and check its fit by placing it in position on top of the wall. This initial rafter is used as a template for marking the remaining three rafters. It's important that all four rafters are exactly the same size. If they're not, you'll have trouble getting the plywood sheathing and roof shingles to fit properly.

Because this roof has a slope of 30 degrees (or a pitch of slightly less than 7-in-12), both ends of each 2×4 rafter are miter-cut to a 30-degree angle. Also, the bottom corner of each rafter is lopped off to make it sit flat on the top wall plate. Although you can determine the length and layout of the rafters mathematically, it is simpler and more accurate to do it with a homemade rafter jig made from three scrap pieces of 1×4. (For more details, see Making a Roof-Rafter Jig on the facing page.)

The quickest, most accurate way to execute these angled cuts is with a power miter saw. You can also use a portable circular saw; just be sure to

Clamp the 1×4 ledger board to sawhorses, then screw on the four 2×4 roof rafters, spacing them 24⅛ in. apart.

clamp the 2×4 to a sawhorse so you can keep both hands on the tool. Use a framing square or sliding bevel square set at a 30-degree angle to accurately mark the cut lines.

Assemble the roof frame

After making all four rafters, cut the 1×4 ledger board to length. The ledger should be the same length as the timber-frame foundation, which is also equal to the width of the shed, as measured from one end wall to the other. The ledger on this shed measures 78⅜ in.

To build the roof assembly, first clamp the ledger board to sawhorses, then mark layout lines for the four 2×4 rafters. Rafters are commonly spaced on 24-in. centers; however, on this simple roof, I positioned one rafter flush at each end of the ledger and equally spaced the other two in between. Attach each rafter to the ledger by driving two 1⅝-in. decking screws through the ledger and into the mitered end of the rafter.

The next step is to set the rafter assembly on top of the shed walls, pressing the ledger flat against the siding and checking to make sure the rafter ends are flush with the edge of the front

MAKING A ROOF-RAFTER JIG

In carpentry, there are many times when the most precise measuring method doesn't require a measuring tape. For example, determining the exact length of a roof rafter for this shed roof is best done with a simple wooden jig. The jig consists of two 18-in.-long pieces of 1×4 and one 8-in.-long 1×4 block. The longer pieces represent the 2×4 rafter and the short block acts as the ledger board. Cut a 30-degree miter in one end of one 18-in.-long board. Screw the 8-in. block to the mitered end of the 1×4. This piece represents the top half of the rafter with the ledger attached.

Make an identical 30-degree miter cut on one end of the second 18-in.-long board. Measure 3⅛ in. along the mitered end and make a mark, as shown in the drawing at right. Hold a combination square against the mitered end and scribe a line across the face of the board at a right angle to the 30-degree miter. Cut along the line to remove the corner from the board. This piece represents the bottom half of the rafter.

To use the jig, hold the two halves together to form a mock rafter. Set it on top of the shed's end wall, with the 8-in.-long ledger block pressed tightly against the siding. Adjust the length of the rafter by sliding the two halves closer together or further apart. When the front edge of the rafter is flush with the front edge of the top wall plate, draw a pencil line along the seam between the two halves.

Screw together the two halves of the jig, using the pencil mark as an alignment guide. Position the jig on the wall again to make sure it fits properly. Unscrew the 8-in. ledger block and use the jig as a template to mark cut lines on a length of 2×4. Cut one rafter and test-fit it on top of the wall; don't forget to slip the 1×4 ledger block behind the rafter. If it fits, mark and cut the remaining rafters.

Roof-Rafter Jig
This simple jig provides an easy way to determine the length of the 2×4 roof rafters.

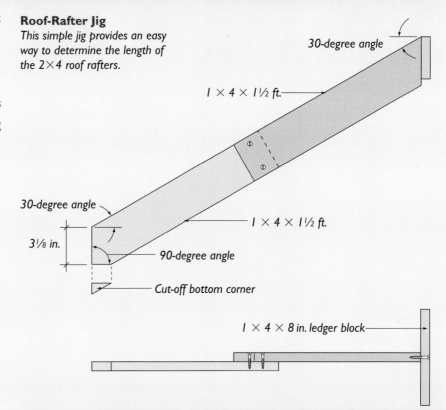

30-degree angle

1 × 4 × 1½ ft.

30-degree angle

1 × 4 × 1½ ft.

3⅛ in.

90-degree angle

Cut-off bottom corner

1 × 4 × 8 in. ledger block

A simple rafter jig can be used to help determine rafter length.

This type of jig will work for nearly any size roof rafter, just be sure to make it from a board that's the same width as the rafter. For example, for a 2×6 rafter, make the jig from a 1×6.

TRADE SECRET

To simplify the installation of plywood siding, try this trick: Drive two nails into the joint between the sole plate and the timber-frame foundation. Leave the nail heads sticking out about ¾ in. Set the plywood panel on top of the nails, which will hold the panel at the correct height. After nailing the plywood in place, yank out the two protruding nails.

WHAT CAN GO WRONG

Be aware that the two pieces of plywood siding for the end walls are similar, but they're not interchangeable. There's a left- and a right-hand piece; they are mirror images of each other. The top edges of both pieces are cut at a 30-degree angle, but they're not identical. To eliminate any confusion, mark cut lines on the grooved faces of each plywood piece, then check to make sure they're marked correctly.

Place a 4-ft. level on top of the ledger board. Attach the ledger to the house wall with 2½-in.-long galvanized decking screws.

Screw the 2×4 in place to provide solid support for the plywood siding. Add a second 2×4 to the right side of the shed.

wall. Be sure to align one end of the ledger board with the outside edge of the end wall by holding a 4-ft. level against the end wall and allowing it to extend past the ledger. Then drive one 2½-in. decking screw through the ledger and into the wall of the house. Move to the other end of the ledger, place a 4-ft. level on top, raise or lower the ledger until it's perfectly level, then screw it to the house. Secure the ledger with 10 more screws, making sure you catch at least two studs inside the house wall. If you screw into only the sheathing behind the siding, the ledger may pull away from the house over time.

With the ledger in place, screw the front end of each rafter to the top of the wall with 2½-in. screws. Finish up the roof framing by measuring and cutting a short vertical 2×4 to fit against the house wall, directly beneath the outside rafters. These pieces of blocking, which are mitered at a 30-degree angle at one end, will provide solid support for the plywood siding. To fasten the 2×4

blocking, drive a screw up through the top wall plate from below, then drive another screw at an angle down through the inside of the rafter and into the top of the 2×4 blocking.

Siding and Trim

The exterior of this shed is covered with ⅜-in.-thick textured plywood siding. Its roughsawn surface has vertical grooves spaced 4 in. apart. I chose plywood siding because it goes up quickly and adds strength and rock-solid rigidity to the 2×4 framing. You'll need three 4-ft. by 8-ft. plywood sheets for this shed: two sheets for the siding and one sheet for the doors.

All the exterior trim is cut from roughsawn red cedar 1×4s. I like using cedar for the trimwork because it takes finish well; is naturally rot resistant; and is dimensionally stable, so it's not likely to twist, warp, or spilt. Despite the shed's compact dimensions, there's quite a bit of cedar trim—nearly 80 linear ft. of trim is installed along the edges of the roof, at all the corners, and around the doorway opening. In addition, there's another 30 linear ft. of trim on the face frames of the two swinging doors.

Install the siding

Before I cut any siding or trim, I generally stain or paint it first. Besides being easier to do ahead of time, this helps protect the wood better. Then I rip one of the full 4-ft. by 8-ft. sheets of ply-wood siding in half for the 24-in.-wide end walls (*ripping* means to cut lengthwise or along the grain, not across it). These pieces won't be exactly 24 in. wide because a little wood is lost to the saw kerf, but the cedar corner boards will hide any gaps.

Next, cut the upper ends of the two plywood end pieces at a 30-degree angle to match the roof slope. Check the fit, making sure the siding extends from the bottom of the sole plate to the top of the rafter. If the panels are a little short, that's fine. The rake trim will cover any spaces. However, the siding mustn't stick up above the rafters, or it'll interfere with the installation of the roof sheathing. When the panel is flush with the wall's sole plate, pressed against the end wall, and tight against the house siding, secure it with 1½-in. ring-shank siding nails spaced about 10 in. apart.

After the two sides have been nailed in place, measure the two front walls and cut the plywood

+ SAFETY FIRST

Take care when handling plywood on windy days. Even relatively small pieces of plywood become impossible to hold onto if they catch a gust of wind. This situation is especially dangerous if you're standing on a ladder. Never set a piece of plywood in place without securing it with at least two nails or screws.

Siding Prep

There are many people who enthusiastically tackle every painting and staining job. They exhibit saintly patience and skillfully apply the finish without ever dripping any on their shoes. I am not one of those people. That's why I prefer to apply finish to plywood siding and cedar trim before installing them. I think that it's much more comfortable working with the pieces spread out across sawhorses: There's no stretching, bending, or climbing up and down ladders. The finish won't run and drip on horizontal surfaces. Plus, you can do a more thorough job because it's very easy to spread an even coat of finish on all edges and surfaces.

For this shed, I applied a medium-gray semitransparent stain on the plywood siding and a light-gray solid-color stain on the cedar trim. A paint roller does a great job of applying stain on the rough-textured surface of plywood siding, but you'll need a narrow paintbrush to get into the deep grooves.

Another option is to buy or rent a paint sprayer. If you go this route, make sure the tool is designed to spray the type of stain you're using. I used a battery-powered garden sprayer, which provided an easy way to stain plywood's rough surface and deep grooves. Note that whenever you spray on a stain, it's very important to back-brush the surface with a stiff-bristle brush to drive the finish deep into the wood grain.

The best tool for painting or staining the 1×4 cedar trim boards is a 3-in.- or 4-in.-wide trim roller. These mini rollers are sold at most paint and hardware stores. Stack the boards in a neat pile and use the roller to coat the edges of the boards. Then take a 1×4 from the pile and roll a coat of finish across the roughsawn surface. When you're done, stand the board off to the side to dry. Repeat the procedure for the remaining boards.

Stain or paint the textured plywood siding before installing it. Use a sprayer or a narrow brush to get finish into the deep grooves.

PRO TIP

Every time you cut a piece of 1×4 trim to length, apply a protective coat of paint or stain to the freshly cut ends.

TRADE SECRET

Here's a sure-fire way to install the front-wall corner boards without any unsightly gaps. Measure the distance from the timber-frame foundation to the fascia. Cut a 1×4 corner board ⅛ in. longer than that measurement. Then use your foot to hold the bottom of the board against the siding. Flex the middle of the board outward and slip the top underneath the fascia. Release the board so it snaps into place, then secure it with siding nails.

Hold the plywood tightly against the house siding and secure it with 1½-in.-long ring-shank siding nails, spacing them 10 in. apart.

Install plywood siding on the front walls. Make sure the siding is wide enough to cover the jack stud at the edge of the doorway opening.

Cover the roof rafters with a piece of ½-in.-thick exterior-grade plywood. Cut it large enough to overlap the plywood siding.

siding to fit. Note that the pieces must be wide enough to overlap the jack studs and tall enough to cover the ends of the rafters. Fasten them to the framing with siding nails spaced on 10-in. centers.

Finally, cut a narrow spacer strip of plywood siding to fit across the header and nail it in place. It doesn't matter whether the grooves on this piece run vertically or horizontally, because the entire strip will eventually be hidden by the cedar fascia board and the doorway header trim.

Sheathe the roof

The next step is to cover the roof rafters with ½-in. exterior-grade plywood (either BCX or CDX grade), which serves as a substrate for the roof shingles. Measure and cut the sheathing to fit flush with the siding on all three sides of the shed, then slide the plywood panel into place, making sure it overlaps the ledger board at the rear. Use 2½-in. (8d) galvanized nails to fasten the sheathing to the rafters. Space the nails about 8 in. apart.

Install rake trim along the sloping ends of the roof. Hold the 1×4 trim flush with the roof sheathing and nail it to the shed wall.

Cut a 1×4 fascia board to fit across the top front of the shed. Make sure it's long enough to cover the ends of the rake trim.

Nail the first corner board to the end wall of the shed. Hold it flush with the plywood siding nailed to the shed's front wall.

Add the trimwork

I've found that it's best to start the trimwork near the roof, then work your way down, saving the doorway trim for last. On this shed, all the trim was cut from red cedar 1×4s and fastened with 1½-in.-long galvanized ring-shank siding nails.

The first pieces to be installed are the angled rake boards that fit along the tops of the end walls. These two pieces follow the roof slope, so you must miter-cut both ends of each board at a 30-degree angle. Cut the rakes to extend down from the house siding and flush with the plywood siding nailed to the front wall. Hold the top edge of each rake even with—or slightly below—the roof sheathing, then fasten it with six 1½-in.-long siding nails. Place two nails near each end and two in the middle.

Next, measure, cut, and install the fascia board, which spans across the front of the shed from one rake board to the other. Before nailing the fascia in place, make sure its bottom edge is even with

the bottom of the rake boards. There will be a small gap above the fascia, but don't worry; it'll be concealed by metal drip-edge flashing.

Now install the corner boards. Trim the two front corners of the shed with two 1×4 boards. Fit one board under the rake and nail it to the end wall; install the other one under the fascia and nail it to the front wall. The two boards overlap to form an L-shaped corner cap. I install the end wall board first so that the second corner board will overlap it and hide the seam from the front.

To follow the pitch of the roof, you'll need to cut a 30-degree miter on one end of an 84-in.-long 1×4. That's a few inches longer than necessary, but it allows you to cut the board to its exact length without having to measure. Hold the corner board in position under the rake board, making sure it's flush with the front edge of the rake. Have a helper make a mark where the corner board intersects the bottom edge of the plywood siding. Then cut the board to length at

TRADE SECRET

After cutting faux-slate shingles to fit along the roof edge, perform a little handi-work so the edges of the shin-gles blend in with the rest of the roof. First, round off the square corner with a utility knife and wood rasp (a very coarse file). Then shave the shingle's just-cut edge with the utility knife to create an uneven bevel. Try to mimic the realistic texture embossed on the opposite, uncut edge.

IN DETAIL

Most roofers apply an asphalt-saturated felt underlayment over plywood sheathing before shingling a house. Besides adding tempo-rary protection for the sheath-ing until the roof is shingled, this thin layer helps protect the shingles from rough and resinous areas on the plywood and offers some protection from wind-driven rain. It's not neces-sary to install felt underlayment on a storage shed. However, make sure the plywood is com-pletely dry before installing the shingles.

After nailing the horizontal head trim to the top of the doorway, install 1×4 vertical trim on each side of the opening.

that mark and secure it with ten 1½-in.-long siding nails. Place two nails near each end and equally space the other six in between.

Measure and square-cut both ends of the second corner board to fit between the timber-frame foundation and the fascia. Position this piece so it overlaps the first corner board, then secure it to the front wall with ten 1½-in.-long siding nails. Place two nails near each end and equally space the other six in between.

The final three pieces of 1×4 trim go around the doorway opening. Start with the horizontal head trim that fits between the fascia and the top of the doorway, cutting it exactly 10 in. longer than the width of the doorway opening. The extra length allows the head trim to extend 1½ in. beyond the vertical 1×4 side trim that will be installed along each side of the opening. You'll also have to rip this 1×4 head trim to fit the narrow space between the doorway and the fascia. Cut the head trim flush with the plywood siding at

the top of the doorway opening. After the head trim has been ripped to width, hold it tightly against the bottom of the fascia, center it in the opening, then nail it in place.

Cut the 1×4 side trim pieces to fit underneath the head trim on each side of the doorway. Align the side trim pieces flush with the edge of the plywood siding, then attach them with eight 1½-in.-long siding nails. Place two nails near each end and equally space the other four in between.

Roofing

Roofing is often the most mundane and tedious part of building a storage shed, but not in this case. First, the roof of this shed locker is only about 20 sq. ft. in total area. You'll be done with the roof long before you ever get tired of nailing down shingles. Plus, it's always a little exciting when you work with an innovative new building product for the very first time.

The Dura Slate shingles used on this shed look like authentic slate but are made of a high-tech polymer that's more like hard rubber. The 12-in. by 18-in. shingles are virtually indestructible, but they are as easy to install as standard asphalt shin-gles. However, before you can begin the installa-tion, you must prepare the roof for the shingles.

Prepare the roof deck

The first step is to install drip-edge flashing around the perimeter of the plywood roof sheathing. The aluminum flashing, which is sometimes called drip cap, prevents water runoff from seeping behind the siding. I installed white drip edge because it matched the white-stained trim. It's also commonly available in brown and unpainted aluminum, which is known as a mill finish.

Start by placing a length of flashing against the front edge of the roof. Hold one end of the flashing even with the outside surface of the rake board, then mark the other end of the flashing where it intersects the opposite rake board. Cut the flashing at the mark with a pair of tinsnips and attach it to the roof with ¾-in. roofing nails spaced 10 in. apart.

Next, install drip-edge flashing along each side of the roof. Cut the pieces so they extend from the house siding to the front edge of the previously installed flashing. Be sure to position the flashing so it overlaps the flashing on the front edge of the roof, then nail it to the plywood sheathing. Use ¾-in. roofing nails spaced 10 in. apart; drive the nails about 1 in. from the rear edge of the flashing.

Install the shingles

Before the main shingling can begin, you must install a starter course of Dura Slate shingles along the three edges of the roof. This initial layer provides support for the roofing and fills in the spaces between the shingles so you won't see the plywood sheathing.

Make a starter shingle by cutting in half a full-size Dura Slate shingle. The simplest way to cut the rubberlike shingle is with a sharp utility knife. Don't try to cut all the way through in one pass; the stuff is much too tough. Instead, score the surface three or four times, using a straightedge to help guide the cut, then bend the shingle until it

Cut the drip-edge flashing to length with tinsnips, then attach it to the plywood roof sheathing with ¾-in.-long roofing nails.

Install drip-edge flashing on both ends of the roof. Cut each piece so it overlaps the flashing nailed along the front of the roof.

snaps in two. Discard the bottom part embossed with the slate texture and keep the smooth top portion.

Begin installing the starter course along the sloping ends of the roof. Align the starter shingles with the edge of the drip-edge flashing and attach them with roofing nails. Then nail the starter shingles along the front edge of the roof, again keeping them flush with the flashing. After the starter course is installed, you can begin shingling the roof.

Draw a vertical centerline down the middle of the roof and use it as a starting point for installing the main shingles. Position a full-size Dura Slate shingle with its right-hand edge on the centerline. Make sure the shingle's bottom edge is even with the drip-edge flashing, then secure it to the ply-

TRADE SECRET

If your stepladder is a little wobbly, check out the steel rods that run underneath each tread. If they have hex nuts on the ends, use a socket wrench or nut driver to tighten them. That will draw together the vertical rails and stiffen the entire ladder. If you can't fix the ladder, don't risk injury; buy a new one.

IN DETAIL

Note that 1×4 exterior trim boards are installed at the two inside corners where the shed meets the house. These trim pieces are optional, but they lend a nice finishing touch to the structure. Miter-cut the upper ends of the trim pieces to a 30-degree angle to match the roof slope. Slip each one into place, press it tightly against the house, then fasten it to the shed with siding nails.

wood sheathing with two roofing nails. Place the nails about 1 in. in from each edge of the shingle and 9 in. up from the butt (bottom) edge. Now work out in both directions from the centerline, laying full-size shingles to the edge of the roof. When you get to the end, hold the last shingle in place and mark where it overhangs the edge.

To cut the last shingle, lay it on a hard, flat surface and use a straightedge to guide you as you score the shingle with a sharp utility knife. Then bend the shingle until it snaps along the scored line, slicing it from the back, if necessary, to separate the two pieces. Align the cut shingle with the bottom edge and end of the roof and attach it with two roofing nails.

To begin installing the second course of shingles, move back to the roof's centerline. This time, center the first shingle directly over the centerline. This offset alignment ensures that the vertical seams between the shingles will be staggered from one course to the next. If the seams aren't staggered, water will flow between the shingles and eventually leak into the shed.

Again, work out from the center of the roof, installing full-size shingles and maintaining a 6-in. exposure to the weather. Trim the end shingles to fit, as described previously. For the third course, return to the centerline and follow the same procedure as the first course: Align the right-hand edge of the first shingle with the centerline. Continue working your way up the roof, trimming the shingles in the final course to fit within ½ in. of the house siding. This particular roof requires five full courses of shingles.

Begin installing starter shingles along the upper end of the roof. Align each shingle with the outer edge of the flashing.

Draw a vertical line down the center of the roof. Align the first shingle with the centerline and attach it with roofing nails.

Hold a full-size roof shingle in place and mark a pencil line along its underside where it overhangs the end of the roof.

Lay the faux-slate shingle on a scrap of plywood. Hold a level on the cut line and score the shingle with a sharp utility knife.

Bend the shingle until it snaps along the scored line. Then carefully slice through the back of the shingle with a utility knife.

Center the first shingle of the next course directly over the centerline and work out in both directions toward each end of the roof.

Install the flashing

A continuous piece of aluminum, lead, or copper flashing must be installed along the seam where the roof meets the house. Without the flashing, water will run down the siding and into the shed. Flashing is available in various widths; I used 8-in.-wide flashing for this shed. You'll need enough flashing to cover the entire width of the roof and ¼ in. on each end. That little bit extra will permit the flashing to overhang the shingles by ⅛ in. on each end of the roof.

Although lead flashing is very malleable, aluminum flashing must be bent lengthwise down the middle to give it a slight L-shape profile. If

PRO TIP

Pay extra for heavy-duty, exterior-grade door hinges. Thin-gauge metal hinges don't adequately support doors with storage racks mounted on the back.

Pry the siding away from the house by about 1 in. or so, then slip a continuous piece of metal flashing behind it.

TRADE SECRET

Buy all the 1×4s for the exterior trim at the same time, making sure you handpick the boards from the same lumber pile. This will greatly increase the chances that all the boards will be exactly the same thickness and width. If you return a week or so later, the new pile of 1×4s may be slightly narrower or thicker. That's because there are often slight size variations between one stack of boards and another.

WHAT CAN GO WRONG

Nothing epitomizes shoddy carpentry more than a pair of ill-fitting doors. Not only do they look terrible, but they also frequently drag and bind. That's why it's important to cut plywood doors perfectly square. Otherwise, they'll look cockeyed when set in the doorway opening. It's equally important to cut plywood panels to exactly the same height. If one's taller than the other, you'll never get them aligned. If the doorway opening is out of square, trim the doors to match it. You'll end up with an even space around each door, and it won't be obvious that they're not square.

you use copper flashing, your metal fabricator can make the bend for you.

Here's an easy way to make the bend in aluminum flashing. First, clamp the flashing to the edge of a plywood sheet so that half of the 8-in.-wide flashing overhangs the plywood. Then set a long 1×4 or 2×4 on top of the flashing, directly above the plywood edge, and press down on the board to crease the flashing.

Before you can install the flashing, you must pull the nails out of the siding course directly above the shed roof. Then, using a flat bar, carefully pry the bottom edge of the siding away from the wall by about 1 in., inserting shims behind the siding to prop it away from the wall. Slip the metal flashing behind the siding and allow its lower edge to extend onto the top shingle course. Remove the shims and nail the siding in place, securing the flashing at the same time. Important: Don't drive any nails through the lower flange of the flashing and into the roof.

Nail the siding back in place, driving the nails through the upper flange of the flashing. Don't nail the lower flange to the roof.

+ SAFETY FIRST

Always wear gloves when cutting and installing metal flashing. The material is very thin and bends easily, but the edges and corners can be razor sharp, especially right after they're cut. Also, never force a length of flashing into place. If your hand slips and slides along the edge, you could receive a deep, serious cut.

If the house is sided with shingles, pull the nails from each individual shingle in the course directly above the ledger. Then slip the flashing behind the shingles and nail them back in place. For a house covered with vinyl or aluminum siding, the job's a bit trickier. Unhook the butt edge of the siding course immediately above the ledger from the course below it. The pros use a siding removal tool—known as a zip tool or zipper—to disengage the butt edge, but a narrow prybar or straight-blade screwdriver works just as well.

After unhooking the siding, slip the flashing behind it. Nail a starter strip—available wherever the siding is sold—along the top of the course directly below the piece of siding you just disengaged. Use 1½-in.-long aluminum nails and space them about 10 in. apart. Drive the nails into the slots in the starter strip and through the flashing, but be careful not to drive the nails too tightly; the strip needs to be able to expand and contract freely. Finish up by hooking the loose siding course to the starter strip.

Doors

This shed features a pair of swinging plywood doors, which are the simplest, quickest type of door to build. It takes about only an hour to build both doors.

Each door consists of a grooved plywood panel trimmed with a 1×4 cedar face frame. The door panel is stained to match the shed's siding, and the face frame has the same light-gray finish as the exterior trim. For extra protection against moisture-related problems, apply a coat of stain to the back and all the edges.

I hung each door with a pair of 8-in. surface-mounted T-hinges, though I could have chosen smaller hinges. In that case, I would have used three hinges per door.

Cut the plywood door panels

Both doors can be cut from a single sheet of grooved plywood. To determine the size of each door panel, first measure the width of the doorway at the top, middle, and bottom of the opening. If everything is square and level, all the measurements should be the same; if there's any discrepancy, choose the smallest dimension. Subtract ½ in. (for clearance) from the width of the opening, then divide by two. The result is the width of each door panel.

To determine the height of the doorway, repeat the same measuring procedure, but this time take the measurements at the left side, middle, and right side of the opening. Then subtract ½ in. to find the height of the door panels. Finally, mark

the door height of 75¾ in. on the plywood and crosscut the sheet with a portable circular saw. Mark the door width on the plywood and cut out the two panels, making each one 19¾ in. wide.

Attach the face frame

The face frame on each door is made from five 1×4 boards: two long vertical stiles and three short horizontal rails. Lay one of the door panels across two sawhorses and cut the two long stiles to the same length as the door panel, then use a caulking gun to apply a bead of construction adhesive to the back of the 1×4 stiles. Clamp each stile to the face of the door panel with three

Apply a zigzag bead of construction adhesive to the back of the long 1×4 stiles that form the face frame of the doors.

Install the intermediate door rail 31¾ in. from the bottom of the door. Fasten the 1×4 rail with construction adhesive and screws.

PRO TIP

Place tubes of construction adhesive in the sun for an hour or so before you use them. The thick glue flows out more easily when it's warm.

TRADE SECRET

Every shed needs some sort of lock, if not for security reasons, then to keep out curious kids. A standard hasp and padlock work fine, but it's easy to misplace the lock. A better option is a lockable hasp, which has its own keyed locking mechanism. Screw the hinged portion of the hasp to the left-hand door. Mount the locking mechanism on the opposite door. Rotate the lock 90 degrees to hold the doors closed. To lock the hasp, use the key.

IN DETAIL

The doors shown here are called inset doors because they fit inside the doorway opening. Overlap doors mount over the doorway, so they're easier to install. This shed required inset doors because the roof doesn't overhang the front wall. If overlay doors were installed, rain would run down behind them.

Center each 8-in. strap hinge over the hinge block. Bore screw pilot holes, then attach the hinges with the screws provided.

Fasten the 2¼-in.-wide plywood astragal strip on the inside edge of the right-hand door. Be sure it overhangs the edge by 1 in.

3-in. spring clamps, then drive 1¼-in.-long decking screws through the rear of the panel.

Next, measure and cut the three 1×4 rails to fit between the stiles, then glue and screw the top and bottom rails flush with the door ends. Position the middle rail so its top edge is 31¾ in. from the bottom end of the door. This places the rail slightly below the midpoint, which makes the doors appear taller. Repeat this procedure to attach the face frame to the second door panel.

Mount the hinges

Each door is installed with a pair of 8-in.-long T-hinges. Before screwing the hinges to the doors, you must first install 3½-in.-wide by 5½-in.-long hinge blocks cut from a cedar 1×4 to support the long hinge leaves. I cut fanciful spear points in the hinge blocks, but you can leave them square or cut a different shape; just make sure the blocks are at least 3½ in. by 5½ in.

Position the center of the lower hinge blocks 10 in. up from the bottom of the door and the upper hinge blocks 8 in. down from the top of the door. Attach each block with construction adhesive and three 1¼-in.-long decking screws driven

through the rear of the door panel. Bore ⅛-in.-dia. pilot holes to prevent the screws from splitting the hinge blocks.

Hold each hinge directly over the center of the hinge block and mark the screw-hole locations. Bore ³⁄₃₂-in.-dia. pilot holes through the hinge block and attach the hinge with the pan-head screws provided.

The right-hand door has an astragal strip attached to its edge. The purpose of this narrow strip is to hold the two doors even when they're both latched shut. Cut a 2¼-in.-wide by 75¾-in.-long astragal strip from a piece of ½-in.-thick exterior-grade plywood. Position the astragal on the inside of the right-hand door, allowing it to overhang the edge by 1 in., then glue and screw the strip to the door.

Hang the doors

Place two ⅜-in.-thick shims on the brick floor in the doorway, then stand one of the doors on top of the shims and tilt it into the opening. The shims will automatically create a ⅛-in. gap at the top of the door and a ⅜-in. clearance space below it. Press the hinge side of the door tightly against

HANGING A DOOR-MOUNTED STORAGE RACK

To get the most out of this compact shed locker, take advantage of every square inch of available storage space. That includes the backs of the doors, which together offer more than 18 sq. ft. of surface area. The challenge is finding the best way to utilize this area without taking up too much space inside the shed. Wire-rack shelving units offer the perfect solution. These rugged steel racks are sold at most home centers and hardware stores in a wide variety of sizes and styles.

For this shed, I bought an eight-tier wire rack measuring 18 in. wide by 72 in. long. Its 5-in.-deep shelves are just right for storing cans, bottles, and small hand tools. To fit the rack onto the back of the left-hand door, I cut off the top shelf with bolt cutters. That reduced the shelving unit to about 64 in. tall. By the way, if you don't have bolt cutters, trim the wire rack with a hacksaw or sabersaw fitted with a metal-cutting blade.

To mount the rack, measure 10 in. down from the top of the door and draw a level line. On the line, screw two plastic mounting hooks (provided with the racks) spaced about 14½ in. apart. Set the shelving unit's top horizontal wire on the hooks and press the rack flat against the back of the door. Check to make sure the shelving unit is centered left to right. Then screw two plastic mounting hooks to the bottommost horizontal wire, but this time place the hooks upside-down to prevent the shelving unit from lifting out of the top hooks.

On the back of the right-hand door, I wanted to hang some larger tools and supplies, so I simply tapped in a few nails. However, I could have installed a wire rack on this door as well.

A wire-rack shelving unit for storing cans and bottles is attached to the left door. Longer items are hung from nails on the right-hand door.

A pair of bolt cutters easily slices through this powder-coated steel-wire rack. You can also use a hacksaw or a sabersaw.

Hang the wire-rack shelving unit on the hooks. Then clip two hooks onto the bottom wire, but this time turn them upside-down.

TRADE SECRET

A hasp is used to lock the two doors together, but you must also install a 3-in. sliding barrel bolt on the inside of the right-hand door to keep the doors from bulging outward. Screw the bolt flush with the top of the door. Then pull the door closed and tap the bottom of the sliding bolt with a hammer to leave an impression on the underside of the header. Open the door and bore a ³⁄₈-in.-dia. by 1-in.-deep hole in the header; be sure to center the bit on the impression mark. Latch the door by sliding the bolt into the hole.

IN DETAIL

The shed's six side shelves aren't fastened to the cleats in any manner. That way, you can easily remove them to store tall, narrow items, such as garden stakes, skis, and fishing rods.

Set the door on top of ³⁄₈-in.-thick shims and tilt it into the opening. Then check for an even gap along the top edge of the door.

After boring screw pilot holes through the mounting holes in the hinge leaves, drive the hinge screws with a cordless drill/driver.

Install the hardware

The right-hand door—the one with the astragal strip—is equipped with a 3-in. sliding barrel bolt that holds the door closed. The bolt slides up into a hole bored in the header. Without this piece of hardware, both doors would sway in and out, even when latched together.

Position the barrel bolt against the astragal strip and flush with the top of the door, then attach it with the short screws provided. Close the door and raise the barrel bolt until it hits the header, then rap the bottom of the bolt with a hammer to leave an impression on the header. Using a drill bit that's the same diameter as the barrel bolt, bore a hole through the center of the impression. Close

the doorway trim, then drive one of the supplied hinge screws through each hinge leaf. Before driving the remaining screws, remove the shims and check to make sure the door swings freely and closes without binding. If necessary, remove one of the hinge screws and tilt the door.

Install the second door in the same manner. Test the swing of both doors, then install the remaining hinge screws.

the door and slide the bolt into the hole. If the fit is too tight, place the drill bit in the hole and wiggle it around a little to enlarge the hole. Finally, measure 36 in. up from the bottom of the door and install a lockable hasp.

Interior Shelves

Inside the shed, tucked around the corner to the left and right sides of the doorway, is a surprising amount of storage space. If left as is, these niches are perfect for standing up tall items, such as pole pruners, post-hole diggers, and garden stakes.

In this storage shed, I decided to utilize the space by installing a few wooden shelves. Another option would be to put up a wire-rack shelving unit, similar to the one mounted on the back of the left-hand door.

Install the side shelves

I made these shelves from 1×12 pine, but cheaper ¾-in. plywood can be used as well. Three shelves are installed on each side of the doorway. Each one rests on a pair of 1×3 cleats mounted to the back and front walls of the shed.

Cut 11-in.-long cleats for the back wall, then fasten them to the corner stud and the perforated hardboard with 1⅝-in.-long screws. Make the front wall cleats 17½ in. long. That way, you'll be able to screw both ends of the cleats to a wall stud. The shelf cleats can be installed at any convenient height; I find it useful to position the lowest shelf 36 in. from the shed floor and then leave spaces of 12 in. and 16 in. between the next two shelves.

Measure and cut the 1×12 shelves so they span the distance between the front and rear walls. Drop the shelves down onto the cleats. You can fasten them to the cleats with screws or just leave them resting in place. There is often enough weight from the items stored on them to keep the shelves from sliding out.

Screw 1×3 cleats onto the shed's rear and front walls to support the side shelves. Use a short torpedo level to align the cleats.

Cut the side shelves from a pine 1×12. Lay them across the cleats. If the shelves fit snugly, you don't need to fasten them to the cleats.

Saltbox Potting

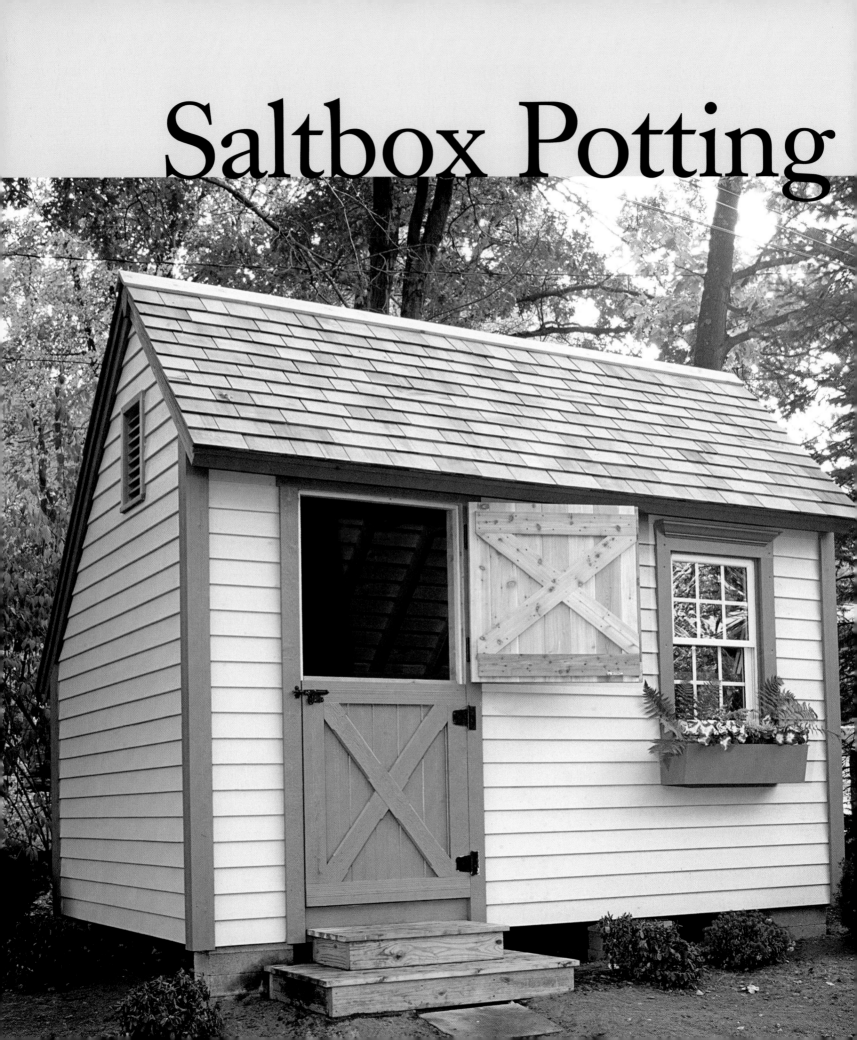

CHAPTER FIVE
Shed

Solid-Block Foundation, p. 108

2 Wall Framing, p. 115

3 Roof Framing, p. 122

Looking more like an enchanted storybook cottage than a backyard storage building, this modestly sized 8-ft. by 12-ft. potting shed features bevel siding, a cross-buck Dutch door, and a traditional saltbox roof covered with red-cedar shingles. At the entrance is an easy-to-build, two-tier platform step. The shed is supported by an on-grade foundation made of eight 4-in.-thick concrete blocks. The walls are framed with 2×4s, then sided with cedar clapboards.

The interior is outfitted with a plywood potting bench that runs along the rear and right-hand end wall. A large perforated hardboard panel is mounted on the wall opposite the bench.

Although it was built as a gardener's potting shed, this pleasing saltbox would also make a great writer's retreat, poolside storage building, or kid's clubhouse. (To order a set of building plans for the Saltbox Potting Shed, see Resources on p. 198.)

4 Roofing, p. 127

5 Dutch Door, p. 130

6 Entry Deck, p. 132

7 Potting Bench, p. 133

PRO TIP

To move solid-concrete foundation blocks to a building site, use a hand truck instead of a wheelbarrow. It's much less likely to tip over.

TRADE SECRET

Measuring the diagonals is a surefire way to determine whether the foundation block layout is square. To make this task much easier, measure both diagonals at the same time with the help of a friend. Stretch the tapes tightly across opposite corners, then read out the dimensions. Whoever has the longest dimension must move the corner block in toward the center. Take another reading and make another adjustment, if necessary. The layout is square when both diagonals are exactly the same length.

WHAT CAN GO WRONG

Buy two or three extra concrete blocks, just in case you accidentally crack a couple while building the on-grade foundation. To reduce the chance of busting a block, make sure you don't set it down on top of a stone or root.

Solid-Block Foundation

The on-grade foundation for this 8-ft. by 12-ft. shed is made of eight 4-in.-thick by 8-in.-wide by 16-in.-long solid-concrete blocks, which are laid in two rows of four blocks each. The blocks in each row are spaced 42 in. on center. The two parallel rows are set 94 in. apart.

This particular building site is elevated slightly above the surrounding yard, so it effectively drains

After removing the grass with a flat shovel, thoroughly compact the soil with a hand tamper. Replace the concrete block.

away rainwater. As a result, the blocks could be set directly on the ground. However, if you have occasional groundwater running across your building site, you must set the blocks on a bed of compacted gravel to prevent erosion.

Set the foundation blocks

After you've determined the location for your shed, begin by setting the four blocks that represent the four corners of the foundation. Remembering to leave at least 3 ft. of air space around the back of the shed, position the first and last blocks in the back row. Space the two blocks 11 ft. 10 in. apart, as measured from the outside end of one block to the outside end of the other. At this point, don't worry about whether or not the blocks are level; just set them in the proper position. Repeat this step for the first and last blocks in the row along the front of the shed, positioning them 7 ft. 10 in. from the ones in the rear row, as measured from the outside edges.

Move each block out of the way and use a flat shovel to remove the grass underneath. Scrape the soil smooth, then pound it flat with a hand tamper. Set the blocks back into place on top of the compacted soil and use a tape measure to confirm that they're back in the correct positions. It's important to build a square foundation so the floor frame will fit properly and be equally supported by every block.

Check the block for level in two directions. If necessary, cover it with a short 2×6 and tap down the high end with a sledgehammer.

Saltbox Potting Shed

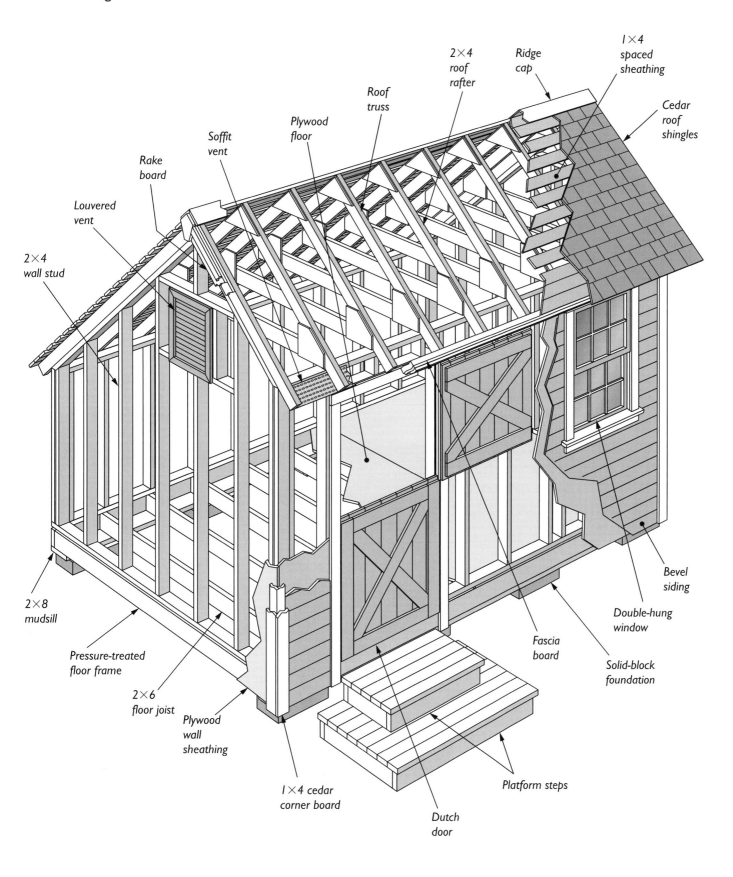

Louvered
vent

Rake
board

Soffit
vent

Plywood
floor

Roof
truss

2×4
roof
rafter

Ridge
cap

1×4
spaced
sheathing

Cedar
roof
shingles

2×4
wall stud

2×8
mudsill

Pressure-treated
floor frame

2×6
floor joist

Plywood
wall
sheathing

1×4 cedar
corner board

Dutch
door

Platform steps

Solid-block
foundation

Double-hung
window

Fascia
board

Bevel
siding

Before nailing down the plywood floor, check the floor frame for square—one last time—by measuring the two corner-to-corner diagonals.

Use a long, perfectly straight 2×4 and a 4-ft. level to check the four blocks for level. Note that the low-lying blocks are shimmed up.

TRADE SECRET

When installing steel-cable ground anchors, it's important to drive the pointed hold-down spike deep into the ground. If the anchors didn't come with a drive pin, substitute a length of ½-in.-dia. rebar. Insert the rebar into the spike, then use a sledgehammer to drive the spike into the ground.

IN DETAIL

For the floor, you'll need three sheets of ¾-in. tongue-and-groove ACX plywood. You could use standard square-edged plywood, but the interlocking tongue-and-groove joints create a much more rigid floor that doesn't bounce.

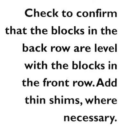

Check to confirm that the blocks in the back row are level with the blocks in the front row. Add thin shims, where necessary.

Using a 2-ft. level, check each block for level in two directions. It may be necessary to move the blocks and scrape away some dirt from any high spots to get the blocks level. The four blocks don't have to be level with each other—yet—but each one must rest on a level, stable spot.

Set a long, straight 2×4 on edge across the two corner blocks along the rear of the shed. Place a 4-ft. level on top of the 2×4. Chances are good that one corner will be lower than the other; in that case, the low corner will need to be shimmed up. You can use additional 4-in.-thick or 2-in.-thick concrete blocks to do this, or if the corner needs to be raised only 1 in. or so, use a 12-in.- to 16-in.-long piece of pressure-treated 1×6 or

strips of asphalt roof shingles. Repeat this procedure to level the corner blocks at the front of the shed.

Next, set the two intermediate blocks between the two corner blocks at the rear of the shed, spacing them 42 in. on center. Again, remove the grass beneath the intermediate blocks, compact the soil, and level each one, then set the long 2×4 and 4-ft. level across the four blocks. If the two intermediate blocks are too low, shim them up; if they're too high, dig out some more dirt from beneath them until the four blocks are perfectly level.

Finally, install the remaining two intermediate blocks between the corner blocks at the front of the shed. Follow the same procedure, first removing the grass beneath the blocks and leveling them side to side, then use a 4-ft. level and a long 2×4 to level the blocks in the front row with the ones in the rear row.

Frame the floor

The structural frame of this shed's floor is made entirely from 2×6 and 2×8 pressure-treated lumber. Start by cutting two 2×8 mudsills and two 2×6 rim joists to 12 ft. long. Next, nail the 2×8s to the 2×6s with 3½-in. (16d) galvanized nails to form two L-shaped assemblies; space the nails about 12 in. apart. Set the assemblies on top of the foundation blocks running across the front and rear of the shed. Make sure the 2×8 mudsills rest on the blocks.

Next, cut ten 2×6 floor joists to 92⅞ in. long and set them between the two rim joists and on

top of the mudsills. It's important that all the joists be cut to exactly the same length and that both ends of each joist be perfectly square. Space the joists 16 in. on center and secure them with three 3½-in. (16d) galvanized nails, driving the nails

Set the 2×6 joists between the two rim joists and on top of the 2×8 mudsills. Fasten each end with three 3½-in. galvanized nails.

Make each mudsill assembly by nailing a 2×8 to a 2×6; cut both boards from pressure-treated lumber rated for ground contact.

Install the remaining floor joists, spacing them 16 in. on center. Nail through the rim joist and into the ends of the floor joists.

ASSEMBLING ROOF TRUSSES

A roof truss is a prefabricated section of the roof frame that consists of two angled rafters and one horizontal bottom chord (ceiling joist). The roof of this 8-ft. by 12-ft. shed is framed with eight trusses spaced 16 in. on center. The rafters and bottom chords are cut from 2×4s. Because this is a saltbox roof, the rafters along the rear of the roof are longer than the rafters at the front.

The three 2×4s that make up each truss are held together with plywood gusset plates that are glued and nailed across the joints on each side of the truss. Therefore, each truss requires six gussets: two at the peak, where the rafters meet, and two at each joint, where the ends of the bottom chord join the rafters.

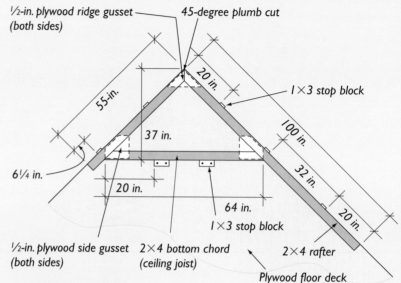

Roof-Truss Assembly
Each roof truss is built from two 2×4 rafters and a 2×4 bottom chord. The parts are assembled on the plywood floor deck. Seven 1×3 stop blocks are used to hold the three boards in position, then plywood gusset plates are glued and nailed across the joints.

Cut the Truss Parts

The quickest, most accurate way to assemble the trusses is to build them directly on the floor deck, using the square corner of the floor to align the rafters. Start by cutting eight 2×4 rafters to 100 in. long and eight more to 55 in. Cut one end of each rafter at a 45-degree angle (known as a plumb cut) and cut the other end square. Cut eight bottom chords to 64 in. long, mitering both ends of each board at a 45-degree angle. Then saw all the gussets out of ½-in. ACX plywood. Cut each of the 16 triangular ridge gussets to 9 in. high by 18 in. wide; cut the 32 side gussets to 10 in. by 10 in.

Install the Stop Blocks

Next, cut seven 8-in.-long stop blocks from 1×3s. Screw three of them vertically to the edge of the floor frame to support the 100-in.-long rafter, then move around the corner and fasten two more blocks to support the shorter rafter. (See the drawing above for the exact placement of the stop

blocks.) Be sure to leave at least 2 in. of the 1×3 stop blocks protruding above the plywood floor deck. You're now ready to assemble the trusses.

Assemble the Truss

On the floor deck, set a pair of rafters in place and press them up against the stop blocks, making sure the 45-degree miters fit together tightly at the peak. Then lay the bottom chord in place, making sure you have the proper overhang of 6¼ in. at the short rafter, and screw the remaining two stop blocks to the floor deck to hold the bottom chord in place. With the seven stop blocks holding the three boards in position, glue and nail on the gusset plates, using construction adhesive and 1-in. roofing nails. Flip over the truss and install the gussets on the other side, then move the completed truss off to one side and repeat the process for the remaining seven trusses. When you're done, don't forget to remove the stop blocks.

Temporarily screw a short block of 1×3 to the edge of the shed's floor frame; it'll act as a stop block while building the roof trusses.

Lay a pair of 2×4 rafters in position on the floor deck. Press them against the 1×3 stop blocks screwed to the edge of the deck.

Lay the 2×4 bottom chord in place, then screw two 1×3 stop blocks to the floor deck to lock the chord in position.

Glue and nail ½-in. plywood gusset plates across the joints. Flip over the truss and install three more gusset plates on the other side.

PRO TIP

Snap chalklines on the plywood wall sheathing to mark the center of each stud. Nail the sheathing along those lines to keep the interior free of sharp nail points.

TRADE SECRET

There's a good chance you'll come across a few twisted or crooked 2×4s. Don't use those boards for the wall studs, wall plates, or roof rafters. Instead, cut them up to use as cripple studs or wall braces. Use the longest, straightest 2×4s for the top and bottom wall plates.

TRADE SECRET

To help position the walls and keep them straight, measure 3½ in. from all the edges of the floor deck and snap chalklines across the plywood before framing the walls. When it comes time to screw the wall to the floor, use the chalklines to help position the bottom plates. If there's a gap between the 2×4 plate and the line, have a helper push in the wall from the outside.

Bore a ⅝-in.-dia. hole in the mudsill, then insert the threaded rod of the ground anchor. Put a large washer and a hex nut on the rod.

Fasten a half-sheet of ¾-in. **ACX** plywood to the floor joists with 2-in. decking screws. Follow up with a full 4-ft. by 8-ft. sheet.

through the front and rear rim joists and into the ends of the floor joists. After the joists are installed, check the floor frame for square by measuring the diagonals. If the two dimensions don't match, use a sledgehammer to lightly tap one corner of the longest dimension toward the center of the frame. When the two diagonal dimensions are equal, the frame is square.

Secure the floor frame to the ground with four steel-cabled ground anchors, which are required by code in some areas. (See According to Code on p. 29.) Bore a ⅝-in.-dia. hole through the mudsill near each corner of the frame, then insert the anchor's threaded rod through the hole and put on a washer and hex nut. Use a steel rod and hammer to drive the hold-down spike into the ground. If necessary, fine-tune the cable's tension by tightening the hex nut on the anchor's threaded rod.

Install the floor

To make the plywood floor, crosscut one sheet of ¾-in. tongue-and-groove ACX plywood in half to create two 4-ft. by 4-ft. pieces. Starting in the front left corner of the floor frame, secure the half sheet to the floor joists with 2-in. decking screws spaced no more than 12 in. apart. Install a full plywood sheet next, making sure the tongue-and-groove joints fit together tightly. If necessary, use a sledgehammer and protective 2×4 block to tap the joint closed. Finally, screw down the two remaining pieces of plywood to complete the floor deck.

Now you can use the flat, clean plywood floor as a giant workbench for building the walls and roof trusses. It is much quicker and easier to build the walls and trusses on a horizontal surface rather than hold up each individual piece. Plus, this method dramatically reduces the amount of time you'll have to spend climbing up and down ladders.

I like to preassemble the eight roof trusses first, then stack them out of the way. (For detailed instructions on building roof trusses, see the sidebar on pp. 112–113.) After that, you can begin building and raising the walls one at a time.

Wall Framing

When professional carpenters build houses, they typically build the walls on the flat floor deck, then tip them up and nail them in place. That same timesaving technique is employed here. Each of the four shed walls is framed, sheathed in plywood, and sided with cedar clapboards while it lies flat on the deck. Even the rake board and gable-end vents are installed ahead of time. When the front wall is set into place, the double-hung window will already be installed and the cedar trim will be nailed around the doorway.

Build a gable-end wall

Begin by building one of the two 8-ft.-wide gable-end walls. These end walls are framed entirely from 2×4s and have a saltbox profile, with the studs at the front of the walls longer than the ones at the rear. As a result, each wall has two top plates that slope up to the peak at a 45-degree angle. When building the end walls, keep in mind that you must also make a rough opening in the frame near the peak to install a louvered gable-end vent.

Build the gable-end wall on the floor deck. Space the 2×4 studs 16 in. on center. Frame a rough opening for a louvered vent.

Cut a pair of 2×4 rafters to length, mitering the upper ends at a 45-degree angle. Attach them to the angled top plates with 3-in. screws.

Cover the wall frame with ½-in. ACX plywood; make sure it extends 7 in. beyond the sole plate at the bottom of the wall.

PRO TIP

For the best, longest lasting protection, apply a coat of stain or primer to both the fronts and the backs of the siding before nailing it to the plywood sheathing.

TRADE SECRET

When screwing the wall sections to the floor deck, position the 3-in. screws about 1 in. from the plywood sheathing. That way, the screws will go through the wall plate and plywood floor and into the 2×6 rim joist. Also, you'll find it much easier to drive the screws if you first bore a ³⁄₁₆-in.-dia. screw-shank clearance hole.

IN DETAIL

To protect the plywood wall sheathing around the louvered vent from water penetration, install strips of felt underlayment. Beginning at the bottom, nail a horizontal piece along the bottom of the opening, then follow it with the two vertical side pieces. Make sure the side strips overlap the bottom piece. Install the top horizontal strip last, making sure it overlays the side strips. Now, if water seeps behind the vent, it won't drip behind the paper.

Cut the five 2×4s needed to form the perimeter of the wall, including the bottom plate, the end studs, and the two angled top plates. Screw the parts together with 3-in. decking screws, then begin cutting and installing the five intermediate studs, spacing them 16 in. on center. When you get to the center of the wall, measure 17¾ in. down from the peak to the top of the opening and frame in a 12-in. by 20-in. opening for the louvered vent. When the rough opening is complete, install the remaining wall studs.

After the wall is framed, cut two 2×4 rafters, making one 100 in. long and the other 55 in. long. These pieces are identical to the rafters you cut earlier for the trusses (see the sidebar on pp. 112–113). Set them on edge on top of the wall's top plates and screw them in place by driving 3-in. decking screws spaced 12 in. apart up through the plate and into the bottom edge of the

Use roofing nails to attach an 8-in.-wide strip of 15-lb. felt underlayment to each end of the wall. Let the felt overhang by 4 in.

rafter. Hold the rafters flush with the outside edge of the plate. Be sure to fasten the longer rafter to the long rear top plate and the smaller rafter to the short front top plate.

Next, prepare to sheathe the wall frame with ½-in. ACX plywood. Lay a full 4-ft. by 8-ft. plywood sheet horizontally across the wall frame, sliding it down so it overhangs the wall's bottom plate by 7 in. Later, when you stand up the wall, the overhanging plywood will conceal the rim joist of the floor frame. Align the ends of the plywood sheet flush with the wall frame, then fasten it with 8d nails spaced 10 in. to 12 in. apart. To square up the end wall before nailing off the sheathing, align the perimeter wall frame with the plywood deck before—and then again after—nailing in the studs.

Cover the remaining wall frame with plywood, trimming it flush with the angled top plates. Lay the plywood over the rough opening for the gable vent, then saw out the plywood to reveal the 12-in. by 20-in. opening. Cut two 3½-in.-wide

Set the louvered vent in the rough opening, then secure it with six screws. Note that the opening is trimmed with felt paper.

strips of plywood and nail them to the rafter tails to bring them flush with the wall sheathing.

Prepare the wall

Before you can begin nailing on the cedar bevel siding, you must first prep the wall. Cut off an 8-ft.-long section from a roll of 15-lb. felt underlayment (builder's paper) and slice the piece into 8-in.-wide strips. Lay one of the strips along both the left- and the right-hand edge of the wall; these are the vertical edges that will form the corners of the building. Allow each strip to overhang the wall by 4 in., then attach it to the plywood sheathing with staples or 1-in. roofing nails. The asphalt-saturated paper will protect the plywood in case any water seeps behind the siding. Once the walls are erected, the paper's overhanging edge will be wrapped around the corner to protect the sheathing behind the corner boards.

Next, cut four 8-in.-wide strips of felt underlayment to fit around the rough opening for the louvered vent. Again, this is necessary to protect the plywood from water penetration. Cut the strips at least 16 in. longer than the width and

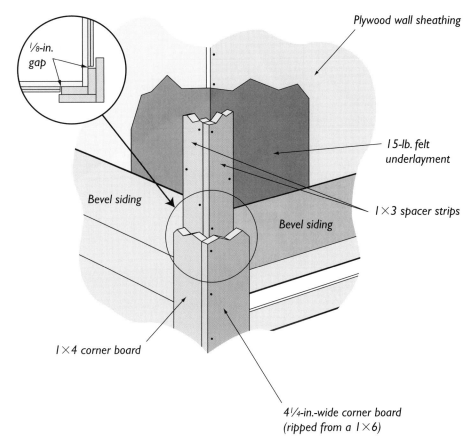

Plywood wall sheathing

¹/₈-in. gap

15-lb. felt underlayment

Bevel siding

1×3 spacer strips

Bevel siding

1×4 corner board

4¹/₄-in.-wide corner board (ripped from a 1×6)

Corner Board Detail
Hidden behind the 1×4 cedar corner boards is a pair of 1×3 spacer strips. The corner boards overlap the ends of the bevel siding and effectively cover the vertical seam between the siding courses and the spacer strips.

Nail a 1×3 spacer strip to each edge of the wall. Drive the nails through the plywood sheathing and into the 2×4 wall frame.

height of the opening, then attach them with staples or 1-in. roofing nails. Run a continuous bead of silicone caulk along the back surface of the louvered vent's frame, set it in the rough opening, and attach it with six brass trim-head screws.

Before you install the siding, nail a long cedar or pressure-treated 1×3 to each edge of the wall, aligning it flush with the outer edge of the plywood sheathing. These 1×3s act as spacer strips and serve a very important function: After the walls are erected, the wider corner boards will overlap the 1×3s and conceal the vertical seam between the siding and the spacer strips. The result is a nice clean corner detail without any unsightly gaps at the ends of the clapboards.

TRADE SECRET

After raising the gable-end wall and screwing it to the floor deck, you must install two diagonal 2×4 braces to prevent the wall from falling over. Make each temporary wall brace from an 8-ft.-long 2×4. Screw one end of the brace to the end wall stud about 5 ft. above the floor deck, then extend the brace diagonally and screw the other end to the rim joist. Fasten each end of the brace with two 3-in. screws. Add a second 2×4 brace on the opposite end of the wall.

Screw a 2×4 brace to the end of the wall and run it diagonally down to the floor frame. When the wall is plumb, secure the brace.

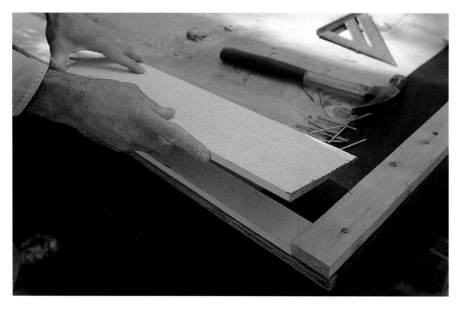

Lay the first course of siding over the starter strip. Let it overhang by ¼ in. to form a drip edge. Tack-nail the siding in place.

Install the siding

Because this shed was going to be painted, I used 6-in. siding with a factory-applied coat of primer. If you're planning to stain your shed, use raw, unprimed clapboards instead. Begin by cutting several pieces of siding to span the distance between the two 1×3 spacer strips. Don't be too concerned about cutting the boards to fit exactly in the space; the corner boards will cover up any small gaps. Next, make a starter strip by ripping a 2-in.-wide piece off the butt edge of one of the clapboards. The purpose of the starter strip is to bump out the first course of siding to match the angle of subsequent courses. Align the starter strip with the bottom of the plywood wall, but don't nail it in place yet.

Next, lay a piece of siding on top of the starter strip, allowing it to overhang the strip by ¼ in. to create a small drip edge. Using 2½-in.-long ring-shank galvanized siding nails, nail through the siding and starter strip and into—but not yet through—the plywood sheathing. If you pound the nails home, the protruding points will prevent you from pressing the wall tightly against the floor frame when it comes time to raise the walls. Install the second course of siding, overlapping the first course by 1 in. to leave 4½ in. of siding exposed

Continue to install the siding, cutting each course to fit between the 1×3 spacer strips. Maintain a 4½-in. exposure to the weather.

to the weather. Again, tack-nail the siding to the plywood, being careful not to drive the nails too far.

When you reach the third course, you can begin driving the nails home. However, be sure to drive the nails into a wall stud to avoid having dozens of sharp points sticking out inside the shed. Continue to cover the wall with siding.

When you come to the louvered vent, cut the siding to fit tightly against each side of the vent's frame. At the peak, trim the siding flush along the edges of the top plates.

Next, use siding nails to attach a cedar 1×4 rake filler strip to each of the rafters at the top of the wall. Miter a 45-degree angle in the ends of the boards where they meet at the peak. Cut both 1×4s about 4 in. too long; the excess will be trimmed off after the fascia is installed. Miter-cut a 45-degree angle in two cedar 1×6 rake boards and nail them on top of each filler board; allow the 1×6s to extend ¾ in. beyond the fillers.

Make a decorative, emerald-shaped keystone block for the peak of the roof. Cut the 4-in.-wide by 5½-in.-high block from 5/4 cedar. Predill the block to prevent it from splitting, then fasten it to the rake boards with three 2½-in. ring-shank galvanized siding nails.

Next, cut two cedar 1×2s to length and set one on top of each 1×6 rake board. Align them flush with the top edges of the 1×6s, butt them tightly against the keystone, then nail them in place. These innocuous little boards, called rake trim, create additional shadow lines along the tops of the gable-end walls, which help emphasize the saltbox's roofline. Again, be sure to run the 1×6s and 1×2s a few inches longer than necessary.

Erect the first gable-end wall

Because these walls are heavy, you'll need at least one person to help you raise them. It's also a good idea to keep all kids and pets away from the site during this phase of the project.

Start by sliding the completed gable-end wall into position. Its bottom plate should be parallel with the end of the plywood floor deck and about 3 in. from the edge. Both you and your helper must grasp the top of the wall by a 2×4 framing member—not the 1×4 rake board—and lift. When the wall is nearly vertical, have your helper

Cover the 2×4 rafters with a 1×6 cedar rake board. Miter-cut the upper ends of the rake at a 45-degree angle where they meet at the peak.

After installing the 1×6 rake board, nail on a 1×2 rake trim. Cut the parts from roughsawn cedar or another weather-resistant wood.

PRO TIP

Each time you crosscut a piece of siding, brush a coat of stain or paint on the fresh end-grain cut to seal out moisture and deter rot.

WHAT CAN GO WRONG

If you're not paying close attention, it's very easy to install clapboards crookedly. Measure each course at both ends, or use a simple homemade gauge block, to ensure that the exposure to the weather remains consistent.

TRADE SECRET

Before nailing the 1×2 fascia trim to the 1×6 fascia, use a handsaw to trim the 1×6 rake board to length, as shown above. Cut through only the rake board, not the 1×2 rake trim. If you can't find rough-sawn cedar 1×6s for the rake or fascia, substitute tongue-and-groove siding. Simply rip the tongue and bevel from one edge and install the board with the grooved edge facing up.

Slowly lift up the completed gable-end wall. Be careful that it doesn't slide off the floor deck, and don't push it beyond vertical.

After the wall's sole plate is screwed to the floor deck, finish hammering in the nails protruding from the first two siding courses.

slide over to the center of the wall while you move around to the other side. Push the very bottom of the wall sheathing tightly against the floor frame, then have your helper on the inside secure the plate to the floor deck with 3-in. decking screws. For now, drive just one screw down through each end of the bottom plate. When the wall is pinned to the floor, hold it upright with two diagonal 2×4 braces.

Finish screwing the bottom wall plate to the floor, starting in the middle of the wall. Have your assistant push in the center of the wall from the outside, then drive a 3-in. screw down through the wall plate. Drive two equally spaced screws between each pair of studs and down through the bottom plate. Step around to the outside of the wall and hammer in the nails sticking out from the first two courses of siding.

Build the remaining walls

The next step is to build and erect the short rear wall, which is only 50¼ in. tall. Start by cutting the 2×4 top and bottom wall plates to 11 ft. 5 in. long. Cut ten 2×4 wall studs to 47¼ in. tall. Next, saw a 45-degree bevel along the top edge of the top wall plate. This is necessary to accommodate the rear rafters that come down from the peak at a 45-degree angle and rest on top of the rear wall.

Screw together the 2×4 wall frame with 3-in. decking screws; set the studs on 16-in. centers. Then install the plywood sheathing, spacer strips, starter strip, and bevel siding just as you did for the gable-end wall. Remove the rear diagonal wall brace from the end wall, then lift the wall into position and secure it by screwing down through the bottom plate. Use a 4-ft. level to check the corner where the rear wall joins the gable-end wall. Push the walls in or out until both are perfectly plumb, then fasten them together by driving five equally spaced 3-in. screws through the rear wall frame and into the end wall.

Now it's time to build another gable-end wall. The second one is identical to the first, except it's a mirror image. Use the same construction techniques as described previously. Don't forget to install a louvered vent and run the cedar rake boards and rake trim a little long. Again, have a helper assist you in standing up the wall. When it's in position, screw it to the floor deck, then check the two walls for plumb in the corner. Attach the rear wall to the end wall with five equally spaced 3-in. screws.

The front wall is the same length as the rear wall—11 ft. 5 in.—but it's 82¼ in. tall. This wall has two rough openings: one for the door and one for the window. Start by cutting the 2×4 top and bottom wall plates to 11 ft. 5 in. long, then cut nine 2×4s to make 79¼-in.-tall wall studs. Begin screwing together the 2×4 wall frame with 3-in. decking screws; space the studs 16 in. on center.

The 34-in.-wide by 74-in.-high rough opening for the door is located 14¾ in. from the left end of the wall. Form the opening with two jack studs, a rough header, and four short cripple studs that fit between the header and the top plate. When framing the rough opening, make sure the 72½-in.-tall jack studs fit underneath the 37-in.-long rough header. Allow the bottom wall plate to continue straight across the doorway. It'll be cut out after the wall is erected.

The 25-in.-wide by 41½-in.-tall rough opening for the window is located 50 in. to the right of the door's rough opening. Just as you did for the door, install a rough header at the top of the window's opening. However, you'll also need to install a rough sill to establish the bottom of the opening. Cut a 5¼-in.-tall cripple stud to fit over the header and a 29½-in.-tall cripple to fit under the sill.

Install the short rear wall along the back edge of the floor deck. Plumb the two walls at the corner and screw them together.

Install the double-hung window in the rough opening in the front wall, then begin nailing on the cedar bevel siding.

Once you've completed the frame, check it for square by measuring the diagonals, then nail the ½-in. ACX plywood sheathing to the framing with 2-in. (6d) nails. Cut 8-in.-wide strips of 15-lb. felt underlayment and nail them to the plywood around the door and window openings. Follow with the same procedure as you did for the louvered vent: Start at the bottom, followed by the two vertical side pieces, and install the top horizontal strip last, making sure it overlaps the side strips. Then carefully set the double-hung

PRO TIP

Conserve battery power when using a cordless drill/driver by rubbing paraffin onto the threads of the screws so they'll go in easier with less friction.

IN DETAIL

Framed into the front wall of this shed are a Dutch door and a 2-ft.-wide window. Because there's nearly 4 ft. of wall space between the window and the door, a much larger window could have been installed. However, the door is an outswinging model and—when open—would have covered part of a larger window. The smaller window sacrifices some natural light inside for a more balanced façade on the outside. When the door is open, the spaces to the right of the window and to the left of the door, and between the window and the door, all measure about 12 in. wide.

TRADE SECRET

Much of the shed's framing is screwed together using a cordless drill/driver and square-drive screws, not Phillips-head screws. Square-drive fasteners, also called Robertson screws, have a square hole, not a cross-shaped slot, milled into their heads; they're driven with a square-drive bit. The main advantage of square-drive screws is that it's virtually impossible for the bit to slip out.

Set the sole plate of the front wall on the floor deck. Tilt the wall back between the two gable-end walls and check it for plumb.

window in the rough opening. Drive a single 2½-in. (8d) galvanized nail through one corner of the window's exterior casing, then start a second nail in the opposite corner. Check to make sure the window is square in the opening and not tilted, then tap the second nail halfway in to hold the window in place. Wait until the wall is raised to finish nailing the window.

Cut three pieces of roughsawn 1×4 cedar to fit around the door opening. Attach the trim boards, called casing, with 2½-in. siding nails, making sure they pass through the plywood sheathing and into the 2×4 wall frame.

Next, install the bevel siding just as you did for the other walls, maintaining a 4½-in. exposure. When you come to the door trim or window casing, cut the siding to fit tightly against each side. Lift the front wall into place, standing it between the two gable-end walls, then secure it by driving 3-in. screws down through the bottom plate and through each corner stud.

After the wall is secured in place, use a handsaw to cut out the section of the bottom wall plate that runs across the doorway. Go over to the window and open and close both sashes to ensure

they slide smoothly and latch securely. Check the window for square by raising the bottom sash about ¼ in. and lowering the upper sash the same amount. Examine the open space between each sash and the window frame. If the window is square, each space will be consistent across its length. If the space is wider at one end than the other, remove the second nail and slip shims between the window jamb and the rough framing. Tap in the shims to adjust the window frame. When it's square, drive the second nail home. Finish attaching the window with a total of six nails, spacing three nails equally along each vertical side casing.

Roof Framing

If you took the time earlier to build the roof trusses before framing the walls, you're about to be rewarded. Framing the shed roof in the traditional manner—cutting and handling individual rafters, collar ties, and a ridge board—can take an entire day. However, it takes only about an hour to set all eight of the prebuilt trusses.

These trusses aren't very heavy, but you'll still need a helper to install them. Set up a stepladder inside the shed so one person can work along the top of the front wall. The second person must hand up the trusses to the first person, then go inside to screw the trusses in place.

Raise the trusses

The trusses are spaced 16 in. on center, but there's no need to measure. All you need to do is set each truss directly over a wall stud. This greatly simplifies the installation process and, more important, helps transfer the weight of the roof to the wall frame and down to the sturdy floor frame and foundation blocks.

Hand the first truss to your helper on top of the wall, then hustle inside the shed and grab the long rear rafter of the first truss. Help your partner set the truss on top of both the front and rear walls, making sure it's directly over the first wall stud. Working at the front wall, align the outside edge of the plywood gusset plate on the truss with the outside edge of the wall's top plate. Drive two 3-in. screws up through the top plate and into the bottom of the truss. Because the stud is directly beneath the truss, you'll have to drive the screws up at a slight angle, placing one screw on each side of the stud. Don't fasten the truss to the rear wall just yet.

Continue installing trusses in this manner, making sure you place one over every stud. Attach the trusses to the front wall only and make sure the edge of each gusset plate lines up with the top plate. After all eight trusses are installed, begin fastening them to the rear wall. However, this time start with the truss in the middle of the roof.

On one of the gable-end walls, measure how far the rear rafter tail overhangs the rear wall, measuring from the end of the rafter to the siding on the rear wall (it should be about 3 in.). Then move to the middle of the shed and have your helper stand outside and measure how far the rafter of the middle truss overhangs the rear wall.

Hand the first truss to a helper at the top of the front wall. Move inside, grab the truss, and set its back rafter on the rear wall.

Secure each truss to the front wall with two 3-in. screws. Drive the screws up through the top plate and into the bottom of the truss.

PRO TIP

Spaced sheathing is ideal for outbuildings because it's easy to install and allows wood shingles to dry out very quickly after they get wet.

IN DETAIL

Aluminum soffit vents come in various widths, ranging from 2 in. to 14 in., and in lengths of up to 12 ft. Some have small slots (little louvers) that allow air to flow through, while others are perforated with thousands of tiny holes; both types work fine. For the Saltbox Potting Shed, you'll need 6-in.-wide vents. If you can't find this size, you'll have to improvise. Go to a lumberyard or building products supplier that caters to aluminum- and vinyl-siding contractors and order 12-in.-wide perforated aluminum soffit vents. Use a utility knife to score the panel lengthwise along the preformed bead that runs down the center of the vent. Carefully bend the panel until it snaps into two 6-in.-wide strips.

Fasten a 1×4 across all the trusses along the peak of the roof. Use 2-in. screws and space the trusses 16 in. on center.

Push in or pull out the wall until the overhang is equal to the one on the gable-end wall; this will ensure that the rear wall is straight. Drive two 3-in. screws up at an angle through the beveled top plate and into the bottom of the rafter. With the middle truss securely holding the rear wall, attach the rear rafters of the remaining seven trusses, checking for a consistent overhang at each rafter tail.

Next, fasten a long pine 1×4 across the rafters at the peak to hold the tops of the trusses in position. Start by measuring the distance between the gable-end walls and cut the 1×4 to match. Hold the 1×4 against the rafters, close to where they're screwed to the front wall. Mark the center

of each rafter on the 1×4. Climb up to the peak and attach the 1×4 to the rafters with 2-in. screws. Hold the top edge of the pine board even with the peak, align each rafter with its centerline, then drive in the screws.

Add soffit vents

The rafter tails extend beyond the front and rear walls by about 6 in., which is just enough to form a small eave. The open space beneath each eave is closed off with a standard 6-in.-wide perforated aluminum soffit vent. The vents allow fresh air to flow into the shed, then exit from the wooden louvered vents mounted on each gable-end wall.

To form each eave, start by cutting a long 2×4 to span the distance across all the rafter tails, including the ones on the gable-end walls. Nail the 2×4—called the subfascia—to the ends of the rafter tails with 3½-in. (16d) nails. Next, slip the 6-in.-wide vent underneath the eave. Hold the vent at an angle, with its upper edge against the very top of the wall and its bottom edge flush with the 2×4 subfascia. Using 1-in. screws spaced 12 in. apart, attach the upper edge of the aluminum vent to the shed wall. Attach the lower

✔ According to Code

Both ends of each roof truss are screwed to the top wall plate. However, the building inspector may ask you to fortify these connections with metal tie-down straps, which are sometimes called hurricane clips. Nail the upper end of the strap to the rafter or the bottom chord, then fasten the lower end to the top plate and the wall stud.

Nail a 2×4 subfascia to the ends of the rafter tails. For easier nailing, first bore ³/₁₆-in.-dia. nail pilot holes.

Fasten the upper edge of the vent to the very top of the shed wall with 1-in. screws. Fasten the vent's bottom edge to the 2×4 subfascia.

Cover the subfascia with a 1×6 cedar fascia board. Fasten the roughsawn 1×6 with 2½-in.-long ring-shank siding nails.

TRADE SECRET

The second course of roof shingles must overhang the starter course by about ¼ in. to form a drip edge. Here's how to establish an even, consistent overhang without measuring. Hold a flat carpenter's pencil against the edge of the starter course. Slide the shingle of the second course until it's flush with the pencil, then nail it in place.

WHAT CAN GO WRONG

If you come across a cedar shingle that's 14 in. wide or wider, slice it in half with a utility knife. This is necessary because very wide shingles have a tendency to bow and curl. Also, it's customary to leave a ¼-in. space between shingles. However, if the shingles are wet, space them ⅛ in. apart instead. When they dry out and shrink, you'll end up with ¼-in. spaces.

Start sheathing the roof by nailing two 1×4s across the rafters at the eave. Rip a 1×4 to 2½ in. and nail it beside the two 1×4s.

edge of the vent to the bottom of the 2×4 with the same size screws.

Next, cut two long pine 1×4s to span the distance across the roof at the eave. Set one 1×4 on top of the rafters, align its lower edge with the outside face of the subfascia nailed across the rafter tails, then fasten it with two 2½-in. (8d) nails at each rafter. Set the second 1×4 on top of the rafters, butt it tightly against the first 1×4, then nail it in place. Use a handsaw to trim the 1×4 rake filler board flush with the outside surface of the subfascia at each end of the eave, being careful not to cut into the back of the 1×6 cedar rake board.

Cut a fascia board from a roughsawn cedar 1×6 and nail it to the subfascia with 2½-in.-long ring-shank galvanized siding nails. Again, use a handsaw to trim the 1×6 rake boards flush with the fascia. Next, nail a 1×2 cedar fascia trim to the fascia. Cut the overhanging end of the 1×2 rake trim even with the fascia trim.

After completing the fascia trim on both roof edges, temporarily screw long 2×4s along the edges of the roof and flush with the top of the 1×2 fascia trim. These boards will act as align-

ment blocks for installing the starter course of cedar shingles.

Install spaced sheathing

The cedar roof shingles on this shed are installed in the traditional manner: over spaced sheathing, not plywood. This centuries-old building technique consists of a series of pine 1×4s nailed horizontally across the rafters. A 1½ -in. space between the boards allows air to circulate behind the shingles. The width of the sheathing boards and the spaces between them are directly related to the amount of shingle exposed to the weather. In this case, the shingles have a 5-in. exposure.

Start by cutting nine 1×4s to span the distance across the roof. Rip one board to 2½ in. wide, set it on the rafters, and slide it tightly against the two 1×4s nailed on earlier. Fasten this board to each rafter with two 2½-in. (8d) nails. The reason you need to have three boards butt together along the eave is to provide adequate nailing support for the first three courses of shingles.

Next, set one of the 1×4 sheathing boards across the rafters. Hold it 1½ in. above the 2½-in.-wide board and nail it in place, again securing it to each rafter with two 2½-in. nails. Install the remaining 1×4s in a similar manner, working your way up the roof to the peak. Now move around to the back of the shed and install spaced sheathing on the rear rafters in the same way.

Roofing

You'll need approximately two squares (200 sq. ft.) of 16-in.-long cedar shingles for this saltbox roof. This includes a little extra to cover any waste. When ordering the shingles, be sure to ask for Number 1 Blue Label shingles, which are cut from all-heartwood timber for optimum decay-resistance.

Fasten each shingle with two—and only two—1½-in.-long galvanized shingle nails. These thin-shank nails are similar to siding nails, but they have

Form the spaced sheathing by nailing a series of 1×4s across the rafters. Leave a 1½-in. space between each board.

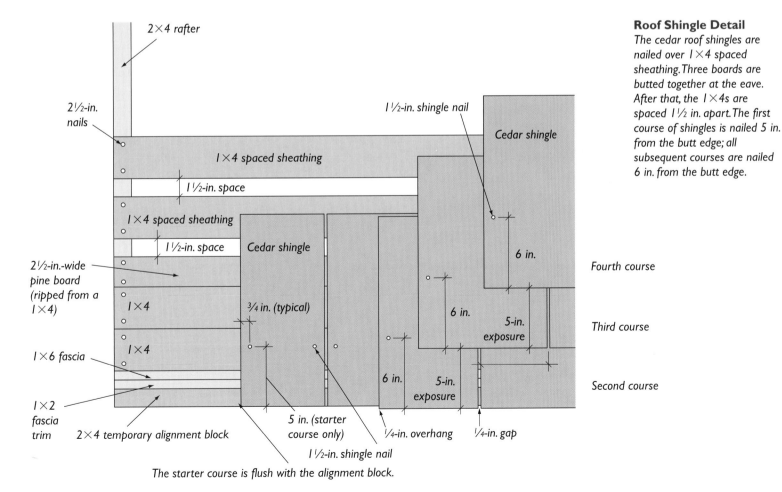

2×4 rafter

2½-in. nails

1×4 spaced sheathing

1½-in. space

1×4 spaced sheathing

1½-in. space

2½-in.-wide pine board (ripped from a 1×4)

1½-in. shingle nail

Cedar shingle

Cedar shingle

¾ in. (typical)

1×4

1×4

6 in.

6 in.

5-in. exposure

Fourth course

1×6 fascia

6 in.

5-in. exposure

Third course

1×2 fascia trim

2×4 temporary alignment block

5 in. (starter course only)

¼-in. overhang

¼-in. gap

Second course

1½-in. shingle nail

The starter course is flush with the alignment block.

Roof Shingle Detail
The cedar roof shingles are nailed over 1×4 spaced sheathing. Three boards are butted together at the eave. After that, the 1×4s are spaced 1½ in. apart. The first course of shingles is nailed 5 in. from the butt edge; all subsequent courses are nailed 6 in. from the butt edge.

PRO TIP

This door's cross-buck face frame is fastened with 1¼-in. nails. If you don't have this size nail, use wire cutters to snip longer ones down to size.

IN DETAIL

Add a little visual interest and style to a Dutch door by routing a chamfer along both edges of all the cross-buck pieces. Use a 45-degree chamfering bit and set it for a ¼-in.-deep cut. Start and stop each cut about 1½ in. from the ends of the boards. Also, be sure to move the router from left to right for a clean, controlled cut.

WHAT CAN GO WRONG

The boards you use to build the doors must be thoroughly dry. If they're even the slightest bit damp, the doors will shrink as the wood dries, resulting in an ill-fitting door. Stack the boards indoors with "stickers" (wood strips) in between to allow air to circulate around each board. Wait at least a week before building the doors.

a slightly larger head. Shingles come in random widths, so be sure to mix and match pieces of various widths as you work your way across the roof. Also, be sure to stagger all vertical seams by at least 1½ in. from one course to the next.

Nail down the shingles

The first course of shingles is commonly called the starter course, though I've heard some carpenters refer to it as the soldier course. Starting at the lower left end of the roof, set the first shingle in place. Hold its butt (bottom) edge even with the 2×4 alignment block and allow its left edge to overhang the rake trim by 1 in. Measure 5 in. up from the butt edge and fasten the shingle with two nails, driving them ¾ in. from the edge of the shingle. Set the next shingle beside the first one, leaving a ¼-in. space between them. Hold the shingle flush with the alignment block, then nail it 5 in. up from the lower edge. Continue installing the starter-course shingles across the roof in this manner. When you reach the end, cut the last shingle to fit with a sharp utility knife, remembering to leave a 1-in. overhang.

The second course of shingles is installed directly on top of the starter course. This time, let the shingles hang down past the starter shingles by approximately ¼ in. That tiny overhang creates an effective drip edge. Measure 6 in. up from the butt edge and drive two nails per shingle. Install the

+ SAFETY FIRST

The roof at the front of this saltbox is relatively small (about 50 sq. ft.), so it's tempting to work on a ladder. However, it's much safer—and more comfortable—to set up a scaffold plank on ladder brackets. Lean two extension ladders against the wall, hook a bracket onto each one, then lay a scaffold plank across the brackets.

rest of the second-course shingles, maintaining the ¼-in. overhang and nailing 6 in. from the butt.

Before installing a third course, measure 5 in. up from the butt edge of the second course and snap a line across the roof. Set the shingles for the third course on this line and nail them 6 in. up from the butt edge. Continue installing shingles in this manner, making sure you always keep a 5-in. exposure to the weather and locating the nails 6 in. up from the butt edge. This way, all the nail heads will be covered by the next course.

As you make your way across and up the roof to the peak, remember to stagger the vertical seams by at least 1½ in. When you reach the last course, you can either cut each shingle flush with the peak before nailing it down or let them all run long, then return later and trim them flush with a utility knife.

At the rear of the shed, use the same techniques to shingle the long back slope of the roof. You'll be able to install the shingles on the lower half of the roof while standing on a ladder or scaffold set up at the eave. However, to reach the upper section, you'll have to install three metal roof brackets and a long scaffold plank about halfway up the roof.

Install a ridge cap

To finish the roof, build a continuous ridge cap from a cedar 1×4 and 1×6. Measure the length of the upper course of shingles at the peak; it should be about 12 ft. 9 in. long. Add 1 in. to that dimension and crosscut the two boards to length. Then use a portable circular saw or a table saw to rip the 1×6 to 4¼ in. wide. Using 2½-in.-long siding nails, fasten the ripped 1×6 to the 1×4 to create an L-shaped cap that's 4¼ in. wide on each side. Set the cap on top of the peak, making sure it overhangs the shingles by ½ in. at each end. Fasten the ridge cap to the roof with 2½-in.-long siding nails, driving two nails through the cap and into every other truss.

Nail in place the starter course of shingles. Hold the butt edges even with the 2×4 alignment block nailed across the rafter tails.

Set the shingles for the third course 5 in. from the second course. Then nail the shingles 6 in. up from the butt edge.

Install a continuous ridge cap along the peak of the roof. Make the cap by nailing together two long roughsawn cedar boards.

Installing Corner Boards

After the roof has been shingled, it's time to apply cedar trim to the four corners of the shed. As with the ridge cap, I find it easier to nail together two boards first to form a preassembled corner cap. Each corner board is made from a roughsawn cedar 1×4 and 1×6. You must rip the 1×6 to 4¼ in. to create a "square" corner.

The top end of the ripped 1×6 must be mitered at a 45-degree angle to match the slope of the roof. To determine the length of this piece, measure from the butt edge of the first course of siding up the end wall to the rake board. For the 1×4, measure from the first course of siding up the front wall to the underside of the aluminum soffit vent, then cut the piece square at both ends. Nail the mitered board to the 1×4 with 2½-in.-long siding nails, spacing the nails 12 in. apart to create the L-shaped corner trim.

At each shed corner is a flap of felt underlayment that was installed before the siding was attached. Bend the flap around the wall corner, smooth any wrinkles, then fasten it with staples or roofing nails. Now slip the corner board into place, making sure it fits tightly against the soffit vent and rake board. Secure it with 2½-in.-long siding nails spaced 12 in. apart.

Fasten the preassembled corner boards to the front wall and the end walls with 2½-in.-long siding nails spaced 12 in. apart.

PRO TIP

To operate a Dutch door as a single unit, install a sliding barrel bolt vertically on the inside surfaces to hold together the two door panels.

IN DETAIL

An outswinging door requires steps that are several inches deeper than ordinary steps. That's because when you open an outswinging door, you must step back to permit the door to swing past.

TRADE SECRET

A Dutch door is built as a single unit, then cut in two. However, this cut is too critical to make freehand, so you must use a straightedge guide. Cut the guide from a rip of plywood with a factory edge, or use a straight 1×6 board. Clamp the guide to the door, making sure it's perfectly parallel with the cut line. The distance from the guide to the cut line must equal the distance from the sawblade to the edge of the saw's shoe.

Use a portable circular saw to cut the batten door into two panels. Note the use of a clamped-in-place straight-edge guide.

Install the upper door panel first. Bore pilot holes, then screw the hinge leaf to the side casing nailed along the doorway.

Slip shims between the upper and lower door panels to create a ¼-in. clearance gap. Screw the lower panel to the side casing.

Dutch Door

Nearly everyone who has seen this saltbox shed has had the same initial reaction: "Wow, I love that door!" There's something about a Dutch door that captures people's attention and makes them smile. Of course, other types of doors can be used, but for this shed I built and installed a Dutch door. (For more information on building a Dutch door, see the sidebar on the facing page.)

Install the door

Because a Dutch door has two panels, the installation is slightly more complicated than for that of a standard door. To help simplify the task, I used surface-mounted hinges and enlisted the help of a friend. You'll need two 4-in. hinges for each door panel; the hinges are screwed to the door panels before the door is installed. I like to position the hinges 4 in. from the top and bottom edges of the panels.

Start with the upper panel, holding it in place at the top of the door opening and pressing it tightly against the right-hand side casing. To ensure that there's proper clearance at the top of the door, slip a ¼-in.-thick shim between it and the 1×4 cedar head casing nailed over the doorway. Bore ³⁄₃₂-in.-dia. pilot holes through the hinge-leaf holes and into the side casing, then attach the upper panel with the weather-resistant screws that come with the hinges.

Install the lower door panel in the same way. First, hold it in place at the bottom of the door opening, using ¼-in.-thick shims between the two door panels to create sufficient clearance. Slide the panel tightly against the right-hand side casing, then attach each hinge leaf with four screws.

On the inside of the door's lower half, install a 4-in. sliding barrel bolt. Place the bolt flush with the top edge of the door and about 5 in. from the latch edge. Mount the mating part of the

BUILDING A DUTCH DOOR

Make the door from V-jointed, tongue-and-groove cedar 1×6s. Lay the roughsawn boards face down on a workbench.

After attaching the four horizontal rails to the back of the door panel, miter-cut the 1×4 pieces to create the cross-buck pattern.

Cut 1×4s to create a cross-buck pattern for the front of each door panel. Nail them in place with their rough-sawn sides facing up.

This Dutch door is basically a batten door, so its construction is straightforward. The entire door is made from V-jointed, tongue-and-groove cedar 1×6 boards. Some of the boards are ripped to 4 in. wide and used to make the vertical stiles, horizontal rails, and X-shaped cross bucks. The door is initially constructed as a single unit, measuring 33½ in. wide by 74 in. high. Then it's cut into two panels, with the bottom half 6 in. taller than the top half.

First, cut seven tongue-and-groove 1×6s to 74 in. long. Then rip the tongue off one board and the groove off another. Lay the boards on a workbench, with their roughsawn surfaces facing down, and join the boards together to form a panel, making sure you start and end with a square-edged board.

Next, cut four horizontal rails from one of the 4-in.-wide cedar boards. Place two of the rails 2 in. from each end of the door, and place the other two side by side across the middle of the door. Adjust them until the seam between these two center rails is 40 in. from the bottom. That's where you'll cut the door into two panels. Measure and cut the 4-in.-wide pieces to create the cross-buck design, then attach them with construction adhesive and 1¼-in. decking screws.

Now it's time to flip over the door and use a circular saw to crosscut it into two panels, using a straightedge clamped to the door as a saw guide. Build a perimeter face frame for each door panel out of 4-in.-wide cedar, setting the vertical stiles between the horizontal rails and attaching each board with construction adhesive and 1¼-in. (3d) nails. Cut and install the X-shaped cross bucks, using the same adhesive and 1¼-in.-long nails.

PRO TIP

Simplify the job of hanging the two door panels by using surface-mounted hinges, which don't require you to cut precise mortises for each hinge leaf.

IN DETAIL

Attach the 4-in.-wide cross bucks to the door with decking screws and construction adhesive; don't use exterior-grade carpenter's glue. The adhesive is very thick, so it won't run from under the boards.

WHAT CAN GO WRONG

A potting bench is installed along the shed's rear wall, directly below the low-slung saltbox roof. If you make the bench too narrow, you won't be able to stand at it without hitting your head against the 2×4 rafters. Cut the plywood for the bench 24 in. deep, which will force you to stand far enough back from the wall to safely clear the rafters.

bolt on the inside of the door's upper half. Engage the bolt when you want to lock together the two halves and operate the door as a single unit.

Entry Deck

The plywood floor of this shed ended up being about 20 in. above the ground, just high enough to accommodate a two-tier wooden platform step. I could have made a set of wooden stairs or even a ramp, but a platform step is more attractive and easier to build. If the shed floor had been about 14 in. above the ground, I would have needed only a single platform.

Each platform consists of a 2×6 pressure-treated frame topped with 5/4 cedar boards, which measure 1⅛ in. thick and 3½ in. wide. If your local lumberyard doesn't carry 5/4 boards, use cedar or pressure-treated 2×4s instead. Build the bottom platform at least 16 in. longer than the width of the doorway opening. Make the top platform equal in length to the width of the doorway and 10 in. shallower than the bottom platform to create a wide, comfortable step.

Set the concrete blocks

To prevent the steps from sinking into the ground, set the bottom platform on top of four 2-in.-thick concrete patio blocks. Smooth the dirt in front of the doorway, then arrange the blocks so that one supports each corner of the platform.

Place a 4-ft. level across the blocks and check for level in two directions: front to back and side to side. If necessary, stack two patio blocks on top of each other to level them.

Build the platforms

The bottom platform measures 27¼ in. deep by 46½ in. wide. Start by cutting two 2×6s to 46½ in. long; these pieces will be the front and back of the frame. Cut four 2×6s to 24¼ in. long to use as the left and right ends and the two intermediate supports.

Screw the front and back pieces to the two ends with 3-in. decking screws. Slide the two support boards between the front and back, space them 16 in. on center, then screw them in place. Set the frame on top of the concrete blocks and check to make sure it extends past the doorway an equal

Make a foundation for the platform steps from 2-in.-thick concrete patio blocks. Check to make sure the blocks are level.

amount on each side. Cut eight 5/4 by 4-in. boards to 48 in. long and nail the boards to the platform with 2½-in. siding nails. Allow each board to overhang the ends of the frame by ¾ in.

Next, build the 2×6 frame for the top platform, which measures 16¾ in. deep by 30½ in. wide. Cut two 2×6s to 30½ in. long for the front and back of the frame and two to 13¾ in. long for the left and right ends, then screw the frame together with 3-in. screws. Now cut five 5/4 by 4-in. boards to 32 in. long and nail them to the platform. Set the top platform on the bottom one and slide the assembly against the shed.

Finally, fasten together the two platforms by driving two 3-in. screws down at an angle through each side of the top platform and into the bottom platform. Also, screw the top platform to the shed's floor frame to prevent the steps from shifting out of position.

Potting Bench

No potting shed is complete without a potting bench, which serves as a gardener's workbench. It provides a convenient place to mix soil, trim plants, and fill pots. You can buy fancy potting benches at most garden shops, but I decided to make one from ¾-in. ACX plywood. The bench is supported 36 in. off the floor by four wooden brackets, which are securely screwed to the wall

Build the platform frame from pressure-treated 2×6s, then cover the surface by nailing on 4-in.-wide cedar boards.

Set the second platform on top of the first. Note that it's considerably smaller to create sufficient room to step up.

Fasten the top platform to the lower one with four 3-in. decking screws.

(Photo © Joseph Truini.)

TRADE SECRET

As an alternative to building a wooden window box, consider installing a metal flowerpot ring. Available at most garden shops, this accessory comes in various configurations and holds from one to four flowerpots. The metal rings are welded to a horizontal mounting strap. On some models, the rings can be pulled opened to hold larger pots or squeezed closed for smaller ones.

TRADE SECRET

Protect the plywood potting bench from water and soil stains by applying three coats of exterior-grade polyurethane varnish. Allow each coat to dry overnight, then lightly sand the surface with 120-grit sandpaper.

Potting Bench Bracket
The plywood potting bench is supported by wooden brackets, which are screwed to the wall studs. Each bracket is made of two horizontal 1×4s and a diagonal 2×4 brace.

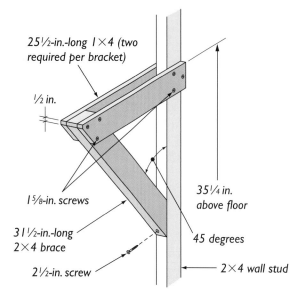

25½-in.-long 1×4 (two required per bracket)

½ in.

1⅝-in. screws

31½-in.-long 2×4 brace

2½-in. screw

35¼ in. above floor

45 degrees

2×4 wall stud

Form each wall bracket by gluing and screwing two 1×4s to a 2×4 diagonal brace. Note that each piece is mitered at a 45-degree angle.

Slide the bracket onto a wall stud. Fasten the 1×4s with 1⅝-in. screws. Secure the 2×4 brace to the stud with a 2½-in. screw.

studs for maximum strength. The bench measures 24 in. wide × 91¾-in. long, but you can make a longer one if you want; just be sure to install more brackets.

Build the brackets

Each bracket consists of two horizontal 1×4s and one diagonal 2×4 brace. Cut the 1×4s to 25½ in. long, mitering one end of each piece at a 45-degree angle. Cut each 2×4 brace to 31½ in. long, mitering both ends at a 45-degree angle. Using carpenter's glue and 1⅝-in. drywall screws, fasten the 1×4s to the 2×4.

Use a 4-ft. level to mark a level line across the rear wall studs 35¼ in. above the shed floor. Slip one of the brackets over the stud, making sure the top edges of the 1×4s are even with the pencil mark. Attach the bracket by driving three 1⅝-in. screws through each 1×4 and into the side of the stud. Also, drive one 2½-in. screw through the

Conceal the exposed edges of the plywood bench with ¼-in.-thick by ¾-in.-wide screen mold. Attach it with glue and brads.

Cut a backsplash from a pine 1×4 and screw it to the studs along the rear of the potting bench. Drive two screws into each stud.

bottom end of the 2×4 brace and into the edge of the stud. Attach the remaining three braces in a similar manner, spacing them equally across the back wall.

Install the bench

Cut the plywood bench to size with a circular saw and set it on top of the brackets. Then fasten the bench to each bracket by driving two 1⅝-in. screws down through the plywood and into the 1×4s below.

To give the bench a more finished look, you can cover the edges with a piece of ¾-in.-wide screen mold, which you can buy at any home center or lumberyard. Apply glue to the front edge and the end of the bench, then nail on the ¼-in.-thick screen mold with ¾-in. brads. When nailing near the end of the screen mold, bore tiny ¹⁄₁₆-in.-dia. pilot holes first, or the brads will split it.

Complete the potting bench by installing a backsplash along the rear wall. This useful accessory will keep items from falling behind the back of the bench. To make the backsplash, cut a pine 1×4 as long as the bench and secure it to each stud with two 1⅝-in. screws.

Storage Idea

On the inside of the front wall, between the window and the door, is a large blank section. Rather than see this perfectly good space go to waste, I converted it into a tool-storage area by installing a perforated hardboard panel (also commonly called pegboard). The panel provides a convenient place to hang various lawn and garden tools. At the top of the perforated panel, I installed two long brackets and a narrow strip of wood to create a shelf for storing clay pots.

The ¼-in.-thick panel measures 43½ in. wide by 54½ in. high; it's attached to the 2×4 wall framing with 1¼-in. drywall screws. Hold the panel against the wall studs, making sure it's tight against the underside of the trusses. Drive the screws into the wall's top plate at the top of the panel and into the studs along the panel's bottom edge and two side edges.

The panel is mounted right beside the doorway and offers more than 16 sq. ft. of space for hanging tools and supplies.

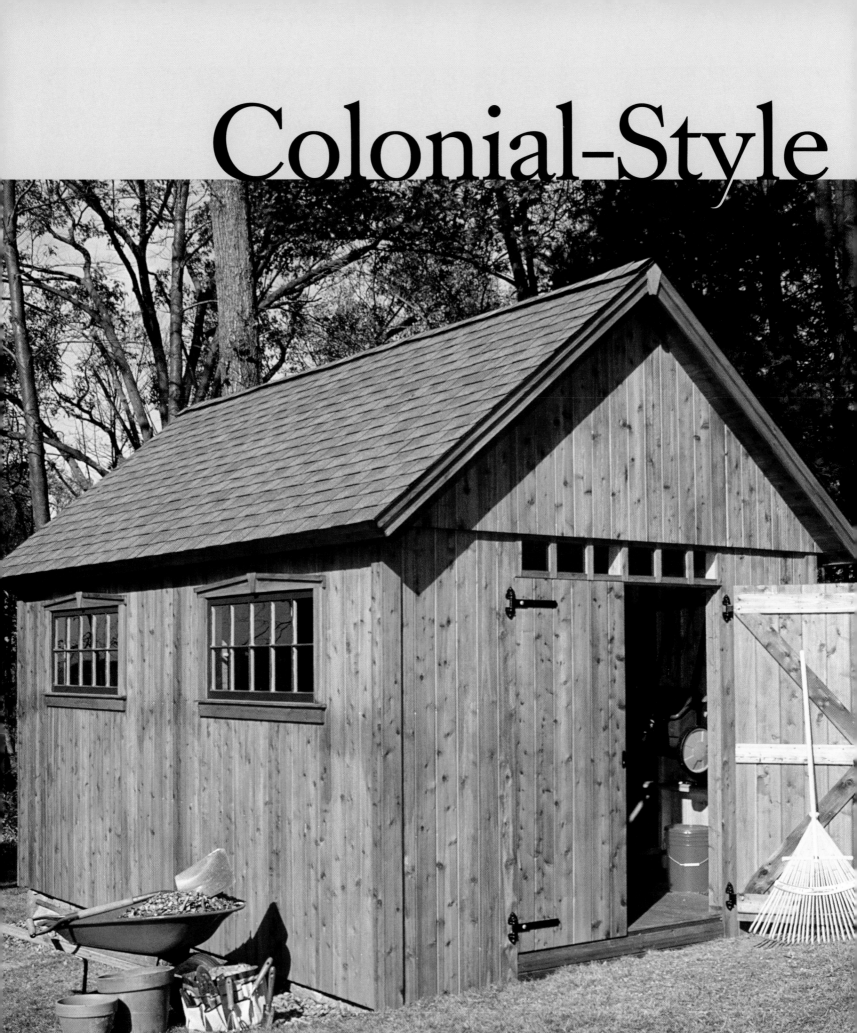

Colonial-Style

CHAPTER SIX
Shed

Skid Foundation, p. 138

Flooring, p. 140

Gable-End Trusses, p. 145

Walls, p. 147

Roof Framing, p. 152

Roofing, p. 156

Windows and Exterior Trim, p. 160

A quintessential storage building, this 10-ft. by 16-ft. colonial-style shed provides more than enough storage space for the average household, yet it is compact enough to fit in the smallest backyard. Because the structure is less than 200 sq. ft., most building departments will allow you to build it on an on-grade foundation. This particular shed was set on a skid foundation formed with 6×6 timbers.

This shed features vertical-board cedar siding, large 2-ft. by 4-ft. barn sash windows, and a pair of double-wide batten doors. The 10-in-12 roof slope is covered with architectural-style asphalt shingles. Note that the roof extends beyond each end wall by about 8 in. to create a gable overhang. This classic architectural detail, which is seldom found on storage sheds, emphasizes the gable roof and creates deep shadow lines at the end walls. (To order a set of building plans for the Colonial-Style Garden Shed, see Resources on p. 198.)

137

TRADE SECRET

Before spreading gravel for a skid foundation, you must first remove the sod. You can rent a sod cutter, but for smaller jobs, a flat shovel works just as well. Use the shovel to cut straight down through the sod all around the area to be removed. Then lift one end of the sod and begin to roll it up. It'll separate easily from the soil and come up in one long strip. Save it for transplanting elsewhere.

WHAT CAN GO WRONG

Here's a common problem that, fortunately, has a quick fix. Sometimes the foundation's skids are level, but when the floor frame is installed, it's slightly out of level. Don't bother messing with the skids; just slip a couple of thin shims under the low end of the frame.

Skid Foundation

This shed features a skid foundation that supports the building on two long, straight timbers laid on the ground in parallel rows. The timbers (skids) are leveled and then the shed's floor frame is fastened to them. For this shed, the skids were made of pressure-treated 6×6s, but 4×6s or 8×8s could have been used as well. It's important to make the skids from pressure-treated lumber rated for either ground-contact use (chemical retention level of .40 lbs./cu. ft.) or wood foundations (.60 lbs./cu. ft.).

Although the skids can be set directly on the ground, it's best to lay them on a bed of gravel for increased stability and long-term support. In most instances, you need to put only a strip of gravel underneath each skid, but in this particular case,

gravel was spread over the entire building site because the area attracts a fair amount of ground-water runoff. The 4-in.-deep gravel bed ensures that the soil beneath the shed won't wash away or remain soggy after a rainstorm. (For more information on skid foundations, see p. 32.)

Set the skids

If you're building the shed on a relatively level lot, you'll need only two 6×6 skids; cut each one to 15 ft. 8 in. long. However, if the site slopes 4 in. or so—as this one did—you'll have to stack two 6×6s to build up the low end of the foundation. To do that, first cut two 6×6s to 10 ft. long for the first (lower) tier of skids, set them in position on the gravel bed, and check them for level. The distance across the skids, from outside edge

Build the skid foundation from pressure-treated 6×6 timbers. When necessary, stack two skids to level a sloped site.

After leveling the skids, counterbore 1-in.-deep holes, then fasten the top 6×6 to the bottom one with 6-in.-long screws.

Colonial Garden Shed

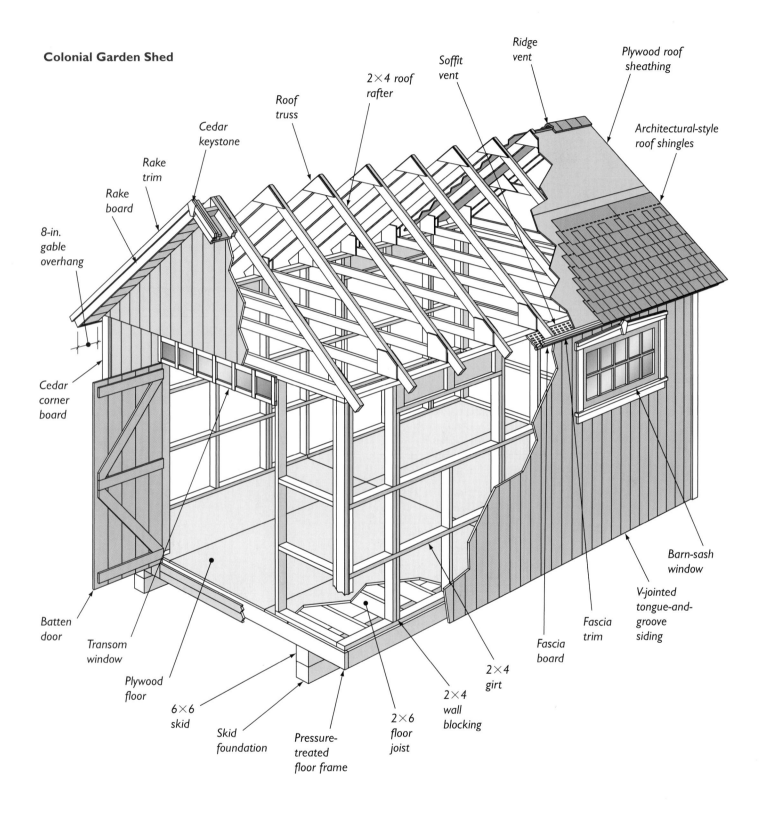

8-in. gable overhang

Rake board

Rake trim

Cedar keystone

Roof truss

2×4 roof rafter

Soffit vent

Ridge vent

Plywood roof sheathing

Architectural-style roof shingles

Cedar corner board

Barn-sash window

V-jointed tongue-and-groove siding

Fascia trim

Fascia board

2×4 girt

2×4 wall blocking

2×6 floor joist

Pressure-treated floor frame

Skid foundation

6×6 skid

Plywood floor

Transom window

Batten door

TRADE SECRET

As you set each joint in place, hold it up and look down its length to see whether it has a slight crown along one edge. If it does, set the board in place with the crown edge bowing upward. That way, it'll eventually straighten out under the weight of the shed floor and the storage items.

IN DETAIL

You'll need five sheets of ¾-in. ACX tongue-and-groove plywood to cover the floor frame. You can save a little money with square-edged plywood, but the tongue-and-groove joints lock together to form a very rigid floor. This extra support is critical for storing tractors, woodworking machinery, or other heavy objects.

to outside edge, must be 7 ft. 4 in. Secure the skids to the ground by first boring three equally spaced ½-in.-dia. holes down through each 6×6. Then use a sledgehammer to pound a ½-in.-dia. by 18-in.-long rebar through each hole and deep into the ground. Be careful not to hit the skid with the sledgehammer, or you may knock it out of level.

Next, cut two 6×6s to 15 ft. 8 in. long and set them on top of the lower-tier skids so the ends and sides of the top-tier timbers are flush with those on the lower tier. Then measure about 1 ft. in from the ends of the upper skids, bore 1-in.-deep holes with a ¾-in.-dia. spade bit, and drive a 6-in.-long screw through each hole to fasten the top-tier 6×6s to the lower ones. Drive four more screws, spaced 24 in. apart, through each upper-tier skid.

Flooring

The understructure frame of the shed's floor is made entirely of pressure-treated 2×6s. It consists of 13 floor joists and two rim joists, which span across the front and rear of the floor frame. Installed on top of the framing is a smooth and strong ¾-in. ACX plywood floor.

Assemble the frame

Start by cutting the two 2×6 pressure-treated rim joists to 15 ft. 8 in. long, then cut the 13 floor joists to 9 ft. 9 in. long. Start assembling the rectangular floor frame by first fastening the rim joists to two of the floor joists, driving three 3-in. decking screws at each corner joint. You'll find it much easier to drive the screws if you first bore ³⁄₁₆-in.-dia. screw-shank clearance holes through the rim joists.

Skid Foundation

When the ground slopes, stack 6×6s to create a level skid foundation. The first tier is partially buried and secured with rebar pins driven into the ground. Full-length skids are set on top and screwed to the bottom-tier skids. For this particular shed, space the skids 7 ft. 4 in. apart.

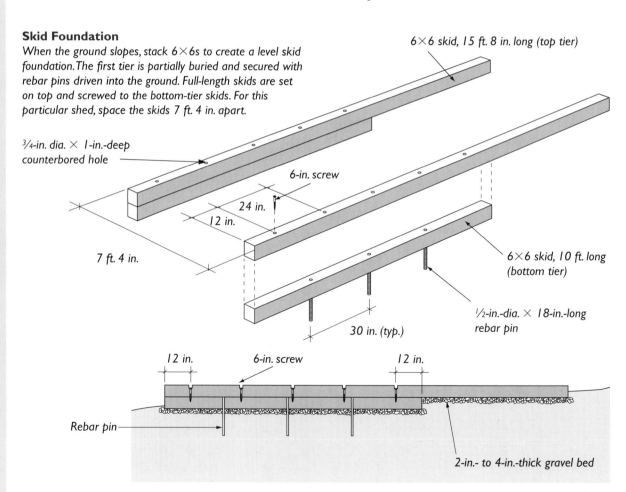

¾-in. dia. × 1-in.-deep counterbored hole

6-in. screw

24 in.

12 in.

7 ft. 4 in.

6×6 skid, 15 ft. 8 in. long (top tier)

6×6 skid, 10 ft. long (bottom tier)

½-in.-dia. × 18-in.-long rebar pin

30 in. (typ.)

12 in.

6-in. screw

12 in.

Rebar pin

2-in.- to 4-in.-thick gravel bed

Screw the perimeter floor frame to the skids with 3-in. screws. Note that the front of the frame overhangs the 6×6 skids by 16 in.

Set the 2×6 floor joists between the front and rear rim joists. Space them 16 in. on center, then fasten the joists with 3½-in. nails.

Next, set the assembled perimeter floor frame on top of the 6×6 skids, with the long rim joists parallel to the skids. The 10-ft.-wide frame is 32 in. wider than the distance between the two skids, so make sure it overhangs each skid evenly by 16 in. Hold the sides of the frame even with the ends of the skids, then drive two 3-in. screws at an angle down through a joist and into one of the skids. With one corner screwed fast, check the frame for square by measuring across the two diagonals. If necessary, tap one corner with a sledgehammer to square up the frame (it's square when the measurements are equal). Finish attaching the frame to the skids with two screws at each of the remaining three corners.

Now set the remaining 11 floor joists between the rim joists and on top of the skids. Install the first 2×6 joist 12 in. on center from the end of the floor frame. Drive three 16d nails through the front and rear rim joists and into each end of the

Secure the floor frame by driving 3-in. decking screws at an angle down through each joist and into the 6×6 skids.

TRADE SECRET

Before screwing the floor frame to the skids, make sure the front rim joist is straight. Stretch a string across the joist, then slip a 2×4 block under the string at each end. Slide another 2×4 behind the string at several places along the joist. Move the frame in or out, as necessary, until the space between it and the string is equal to the thickness of the block.

IN DETAIL

There are only two barn sash windows in this shed, even though, at 10 ft. by 16 ft., the shed could easily accommodate three or four more windows. However, because it was built as a storage building, no windows were installed in the rear wall or the right-hand sidewall. That leaves plenty of wall space to hang tools and install shelves.

joist. Then install the remaining 10 joists, spacing them 16 in. on center.

After all the joists are nailed in place, check the frame for square one last time before fastening each joist to each skid with 3-in. screws. Drive the screws at an angle down through the joists and into the skids. If you find that the screws are splitting the joists, bore 3⁄16-in.-dia. screw-shank clearance holes.

If the local building code requires that the shed be secured with ground anchors, install them now, before nailing down the plywood floor. Bolt one anchor to a floor joist near each corner of the frame. Use a long steel pin to drive the pointed hold-down spike deep into the ground.

Install the plywood floor

Start by installing a full 4-ft. by 8-ft. sheet of 3⁄4-in. ACX tongue-and-groove plywood along the rear edge of the floor frame. Align the sheet perpendicular to the floor joists, with its grooved edge flush with the rear rim joist. Fasten the sheet to the floor joists with 8d galvanized nails spaced about 12 in. apart. Then crosscut the next sheet to 92 in. long, set it end to end with the first sheet, and nail it down. Make sure the end seam falls on the center of a joist.

Crosscut a full-size sheet in half to form two 4-ft. by 4-ft. pieces. Cut one of those pieces to 44 in. and lay it edge to edge with the installed 92-in. sheet. If you have trouble closing the tongue-and-groove joint between the two sheets, gently tap them together with a sledgehammer and a protective 2×4 block. Nail the 44-in. piece to the joists, then install a full sheet, followed by the 4-ft. by 4-ft. half sheet. These three pieces add up to 15 ft. 8 in., the exact length of the floor frame.

Rip the last full sheet of plywood lengthwise down the center to create two 2-ft. by 8-ft. strips. Crosscut one strip to 92 in. Finish covering the floor by nailing the two plywood strips to the joists along the front edge of the floor frame.

Before building and erecting each of the shed's walls, I use the large, clean plywood floor surface

Use a protective 2×4 block and a sledgehammer to tap together two sheets of tongue-and-groove plywood floor decking.

to lay out and build the seven roof trusses. (For more detailed instructions, see the sidebar on pp. 144–145.) It's much quicker and easier to build the walls on the floor deck, then tip them up into place.

Gable-End Trusses

The two gable-end trusses are very similar to the standard roof trusses (see p.144). As with the other roof trusses, each one is assembled with two 87-in.-long rafters and a 10-ft.-long bottom chord. However, the gable-end trusses have two additional 2×4 structural supports, which are attached with plywood gusset plates.

After fastening siding to the trusses, build the rake overhangs before setting the assembly in place on the end walls. This entire process is done on the floor, making it a lot easier and safer than working on a ladder.

Install structural supports

The first structural support is a horizontal collar tie installed between the rafters 19 in. above the bottom chord. This 76¾-in.-long board adds rigidity to the truss and provides nailing support for the V-jointed cedar siding.

The second structural support is an 8-ft.-long shoe plate fastened flush with the bottom chord. When the truss is installed, the shoe plate rests on top of—and is screwed to—the top plate of the gable-end wall. When fastening the structural sup-

Cover the gable-end truss with V-jointed cedar siding. Secure the 1×6 tongue-and-groove boards with 1½-in. siding nails.

ports to the gable-end trusses, be sure to nail the plywood gusset plates on one side only—the side that faces in toward the center of the shed.

Form the gable overhang

After assembling the trusses, cover their exterior surfaces with V-jointed, tongue-and-groove cedar 1×6s. For now, let the boards run long at the top of the truss. Cut the siding to overhang the bottom chord by 1 in. Nail the siding to the rafters, bottom chord, and collar tie with 1½-in. ring-shanked galvanized siding nails, driving the nails at an angle down through the tongue of each board. Using a circular saw, trim the cedar boards flush with the top of the rafters.

Next, build the gable overhang on each of the gable-end trusses. This overhang, also called a rake overhang, is essentially an 8-in.-deep eave framed with 2×4s. It's attached to the top of the truss, then its underside is covered with short pieces of 1×6 cedar siding to form a soffit.

Start by cutting four 87-in.-long rafters from 2×4 stock. Make a 40-degree miter on one end of each rafter. Cut 10 short 2×4 blocks to 5 in.

BUILDING ROOF TRUSSES

Each roof truss is made from two angled 2×4 rafters and one horizontal 2×4 bottom chord (ceiling joist). The three parts are held together with ½-in.-thick plywood gusset plates glued and nailed across the joints. This particular roof is framed with seven standard trusses, which are spaced 24 in. on center, and two gable-end trusses. These last two trusses are installed atop the end walls of the shed and, just like the walls, are sided with V-jointed, tongue-and-groove cedar 1×6s.

It's best to assemble the trusses right on the floor deck. However, because of this shed's roof pitch, you can't use the square 90-degree corner of the floor to

Temporarily screw two 2×4 roof rafters to the ply-wood floor with 3-in. screws. Make sure the joint at the peak fits together tightly.

align the rafters, as you did for the Saltbox Potting Shed's roof trusses (see Chapter 5). That's because the rafters for this shed are cut at a 40-degree angle; when assembled, they form an 80-degree angle. But there's an alternative assembly method that's just as easy and accurate.

Start by cutting the truss parts to size. For each truss, cut two 2×4 rafters to 87 in. long. Cut a 40-degree miter on one end of each rafter; cut the other end square. Next, cut a 2×4 bottom chord to 120 in. long, mitering both ends at a 50-degree angle. Then saw the gusset plates from ½-in. ACX plywood. Cut each of the triangular ridge gussets to 6¾ in. high by 17 in. wide. Cut the side gussets to 11 in. square, then lop off the corner, as shown in the drawing at left. Also, cut eight 4-in. by 4-in. plywood stop blocks.

In the center of the floor deck, lay out two rafters and a bottom chord to form one roof truss, tem-porarily screwing the parts to the floor with 3-in. screws. After making sure the joints between the boards fit tightly together, use 1⅝-in. screws to fasten

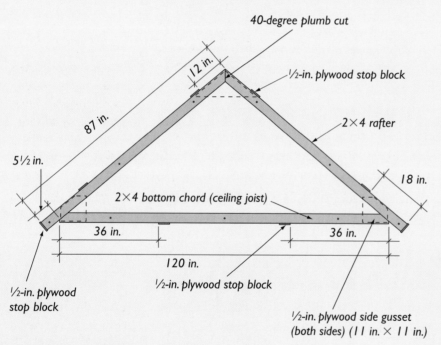

40-degree plumb cut

12 in.

½-in. plywood stop block

87 in.

2×4 rafter

5½ in.

18 in.

2×4 bottom chord (ceiling joist)

36 in.

36 in.

120 in.

½-in. plywood stop block

½-in. plywood stop block

½-in. plywood side gusset (both sides) (11 in. × 11 in.)

Roof-Truss Assembly
Each of the seven roof trusses is built from two 2×4 rafters and a 2×4 bottom chord. One truss is screwed to the plywood floor deck and used as a template for assembling the remaining trusses. Eight plywood stop blocks are screwed to the template truss; they hold the rafters and bottom chord in proper alignment. Plywood gusset plates are glued and nailed across the joints to hold each truss together.

the plywood stop blocks to the outside edges of the bottom chord and rafters. You'll also need to attach a stop block to the end of each rafter tail.

Now use this setup as a template for assembling the seven standard roof trusses. Set one pair of rafters on top of the rafters screwed to the floor deck. Press them against the stop blocks and make sure the 40-degree miters fit together tightly at the peak. Lay a bottom chord on top of the one screwed

to the floor. With the eight stop blocks holding everything in position, glue and nail on the gusset plates, using construction adhesive and 1-in. roofing nails. Flip over the truss and attach gussets to the other side. Remove the completed truss and repeat the process for the remaining trusses. When you're done, remove the stop blocks and fasten gusset plates to the truss screwed to the floor deck. Then unscrew the truss, flip it over, and install gussets on the other side.

Use 1⅝-in. screws to fasten 4-in. by 4-in. plywood stop blocks to the outside edges of the rafters and bottom chord.

Complete the roof-truss template by screwing one ½-in.-thick plywood stop block to the end of each rafter tail.

Lay two rafters and a bottom chord on top of the truss template screwed to the floor deck. Make sure the three parts fit together tightly.

Attach the triangular plywood ridge gusset to the rafters with construction adhesive and several 1-in.-long roofing nails.

PRO TIP

Preassemble the roof trusses on the floor deck—you'll reduce the roof-framing time from an entire day to just a couple of hours.

TRADE SECRET

After assembling the roof trusses, set them off to one side on top of a few scrap 2×4s; don't lay them right on the ground. Cover the stack with a waterproof tarp. If the trusses get wet, they'll be much heavier to lift into place.

(Photo © Joseph Truini.)

IN DETAIL

Larger sheds often have a blank wall section that needs a little dressing up. An easy way to add visual interest to a façade is to install a trellis, then plant a climbing vine or creeping flower at its base. The trellis shown here is made of a prefabricated lattice panel sandwiched between vertical 1×2s. Never attach a trellis flat against the shed; always leave a 2-in. space behind it so the plant has room to grow.

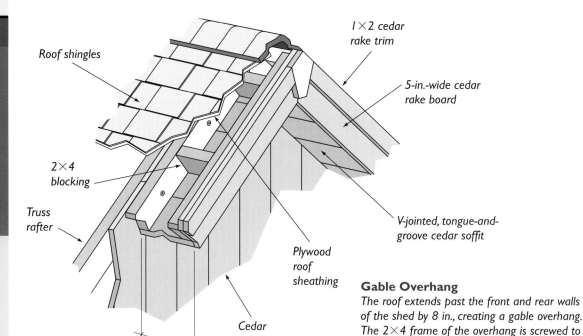

Labels on diagram:
Roof shingles
1×2 cedar rake trim
5-in.-wide cedar rake board
2×4 blocking
Truss rafter
V-jointed, tongue-and-groove cedar soffit
Plywood roof sheathing
Cedar siding
8 in.

Gable Overhang
The roof extends past the front and rear walls of the shed by 8 in., creating a gable overhang. The 2×4 frame of the overhang is screwed to the gable-end truss. Its underside is covered with short pieces of cedar siding to form a soffit.

long. Stand five of the short blocks between two of the rafters, spacing them equally, then screw them together with 2½-in. decking screws to form a ladder. Assemble the remaining two rafters and five blocks to build the second frame. Attach the frames to the top of the truss with 3-in. decking screws. Drive the screws down through the frame and into the rafter.

Form a soffit on the underside of the gable overhang with short pieces of V-jointed cedar; secure each piece with two nails.

Cut 8-in.-long pieces of V-jointed, tongue-and-groove cedar 1×6s—enough to cover the entire underside of the overhang—and fasten them to the underside of the overhang with 1½-in. siding nails. These pieces form the soffit of the gable overhang. Secure each board with two nails driven through the tongue edge.

Cover the face of the overhang—and the ends of the soffit boards—with a 5-in.-wide rake board ripped from a cedar 1×6. Position the rake so it projects ½ in. above the top edge of the 2×4 overhang; the raised lip will hide the edge of the plywood roof sheathing. Next, make a decorative emerald-shaped keystone block for the peak of the roof. Cut the 4-in.-wide by 5½-in.-tall block from 5/4 cedar. Predill the block to prevent it from splitting, and fasten it to the rake boards with three 2½-in. ring-shank galvanized siding nails. Nail a 1×2 rake trim piece to the rake board, keeping it flush with the top edge of the rake board and butted tightly against the keystone. Cut both the rake board and the trim piece about 6 in. too long, then trim them to size after the fascia is installed.

Walls

The walls of this shed are built on the floor deck, then tipped up and nailed into place, a timesaving technique practiced by many professional home-builders. Each wall is framed with 2×4s and sided with V-jointed cedar boards while it lies flat on the deck. The window and door openings are framed in the walls, so there's no additional framing necessary after the walls are erected.

Start by building the left-hand sidewall out of 2×4s. This wall has the two 2-ft. by 4-ft. barn sash windows; the opposite sidewall has no windows. Note in the drawing below that the wall is framed with both vertical blocking and horizontal girts (or purlins). The pieces of blocking serve as wall studs to support the weight of the roof. They're spaced 24 in. on center and are cut to fit between the long horizontal girts, which span the entire length of the wall. The girts provide solid support for nailing up the vertical cedar siding; they also help keep the wall straight. Therefore, it's imperative that you use only perfectly straight 2×4s for the girts.

Frame the sidewall

Begin by cutting the 2×4 top and bottom wall plates to 15 ft. 8 in. long. Cut two end studs to 79¼ in. long, then fasten the wall plates to the end studs with 16d nails to create a rectangular wall frame. Rip one 3½-in.-wide strip of ½-in. plywood to fit against the inside surface of each end stud; attach the strips with 1¼-in. screws. After the wall is complete, these plywood spacer strips help create a sturdy 3½-in.-wide corner post at each

Sidewall Framing Plan
The shed walls are framed entirely out of 2×4s. Long, straight 2×4 girts help keep the wall straight and provide solid nailing for attaching the vertical-board siding. Vertical blocking is installed between the girts to strengthen the wall and support the roof.

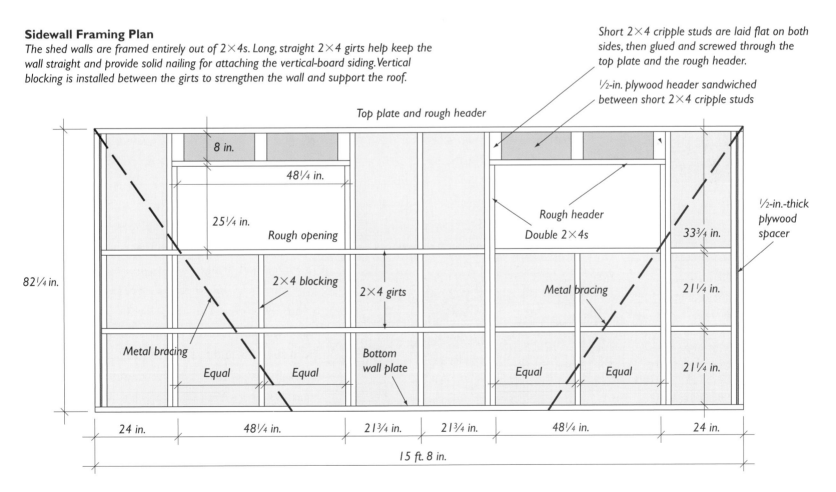

Short 2×4 cripple studs are laid flat on both sides, then glued and screwed through the top plate and the rough header.

½-in. plywood header sandwiched between short 2×4 cripple studs

Top plate and rough header

8 in.

48¼ in.

25¼ in.

Rough opening

Rough header

Double 2×4s

½-in.-thick plywood spacer

33¾ in.

82¼ in.

2×4 blocking

2×4 girts

Metal bracing

21¼ in.

Metal bracing

Bottom wall plate

Equal   Equal

Equal   Equal

21¼ in.

24 in.   48¼ in.   21¾ in.   21¾ in.   48¼ in.   24 in.

15 ft. 8 in.

PRO TIP

There are two types of galvanized nails: electroplated and hot-dipped. Electroplated nails are smoother, but hot-dipped nails have a thicker coating.

TRADE SECRET

You can cut all the framing for a shed like this one with a circular saw, but it won't be easy. There are nearly three dozen parts in a single sidewall; that's over 70 square crosscuts per side. Instead, use a 10-in. power miter saw.

IN DETAIL

When building a wall with vertical blocking and vertical-board siding, you must install diagonal bracing to strengthen the wall frame. The simplest type to install is metal let-in bracing, sometimes called wind bracing. Snap two diagonal chalklines from the center of the bottom plate to a point on each end stud about 4 in. down from the top wall plate. Then two diagonal cuts (about 1 in. deep) along the lines. Tap the metal bracing into the kerf and secure it with roofing nails.

end of the wall. Check the wall frame for square by measuring the opposing diagonals (when the two dimensions are equal, the frame is square).

Next, cut 18 pieces of 2×4 blocking to 21¼ in. long. Stand nine of the pieces along the bottom wall plate, spacing them 24 in. on center, and fasten each one with two 16d nails driven up through the bottom plate. Cut two 2×4 girts to 15 ft. 4 in. long (remember to use lumber that is as straight as possible). Set one of them between the end studs and on top of the nine pieces of blocking fastened to the bottom plate, and fasten the girt to each 2×4 block with two 16d nails. (See the drawing on p. 147.)

Stand the remaining nine pieces of blocking on top of the first girt, making sure you align them with the first row of blocking. Toenail each block to the girt with two 8d nails, then install the remaining girt on top of the second row of blocking, nailing it in place with 16d nails. Cut nine more pieces of blocking to 33¾ in. long and nail these final blocks between the second girt and the top wall plate.

Refer to the drawing on p. 147 for the location of the rough openings for the two 2-ft. by 4-ft. windows. Note that a header (see the drawing on p. 147 for construction details) is formed over each opening with six 2×4 cripple studs and an 8-in.-wide strip of ½-in.-thick plywood. Three

+ SAFETY FIRST

When it comes time to raise the assembled walls, be sure to enlist the help of at least three able-bodied friends. The sidewalls, which are nearly 16 ft. long, are particularly unwieldy and tricky to lift into position. When lifting them, remember to keep your back straight and lift with your legs. Also, don't attempt to raise the walls on a windy day. A gust could blow over a wall, possibly injuring someone.

The sidewall is framed with pieces of vertical blocking and long horizontal girts; there are no continuous wall studs.

cripples are laid flat on each side of the plywood web and fastened with 3½-in. (16d) nails.

Finish the wall

Once the wall frame is complete and checked again for square, install two pieces of diagonal metal wind bracing, which will strengthen the wall and help hold it square. (For more information, see In Detail at left.) Now begin installing the V-jointed, tongue-and-groove cedar siding. Cut the 1×6 siding boards to 89¼ in. (7 in. longer than the wall height). Later, when you stand up the wall, the overhanging siding will cover the 2×6 joist of the floor frame.

Start at the left end of the wall and set the first board flush with the top wall plate and even with the end stud. Make sure the tongue edge faces toward the center of the wall. Fasten the left side

Install the second row of 2×4 vertical blocking directly over the first row. Then slide the next horizontal girt into place.

Create a header over the window opening by sandwiching a ½-in.-thick plywood panel between six 2×4 cripple studs.

Cover the exterior surface of the wall with V-jointed cedar siding. Note that the boards extend 7 in. past the bottom wall plate.

of the board (the grooved edge) by nailing through its face with 1½-in. siding nails, driving the nails into the horizontal framing members (i.e., the top and bottom wall plates and girts). Fasten the other side of the board by toenailing at an angle down through the tongue.

Set the next board in place alongside the first one, making sure the tongue-and-groove joint fits together tightly. Secure it by nailing through the tongue and into the 2×4 framing. (You only need to face-nail the first and last boards.) Install the remaining boards in a similar fashion, checking occasionally to make sure the siding isn't tilted out of plumb.

When you come to the window openings, cut the siding to fit from the rough sill and past the bottom plate by 7 in. After filling in the wall section below the window opening, cut short pieces

Keeping Vertical Siding Vertical

When installing vertical siding, it's a good idea to stop occasionally and make sure the boards are running true and not tilted out of plumb. To do this, measure the distance across the top of the wall to the outside edge of the third cedar board, then take another measurement along the bottom of the wall to the board's edge. The two measurements should be the same. If they're not, make up the difference a little at a time over the next few boards.

For example, if the measurement along the top of the wall is ⅜ in. greater than the one at the bottom, leave a ⅛-in. gap between each of the next three boards along the bottom of the wall. By making up the discrepancy over a span of three or four boards, the adjustment will be virtually undetectable.

PRO TIP

Use a 20-oz. or heavier framing hammer to build the walls. You'll be able to drive larger nails with fewer blows than you would with a 16-oz. hammer.

TRADE SECRET

Here's an easy way to increase the cutting capacity of a power miter saw. Lay a scrap 2×6 across the saw table, then place the board you're cutting on top. That raises the board to the widest part of the blade, which increases the cutting capacity of a 10-in. miter saw from about 5½ in. to 6½ in.

IN DETAIL

This 10-ft. by 16-ft. shed is large enough to be split into two distinct storage areas. Here, a partition wall was built 4 ft. inside the gable-end doors. The wall is covered with perforated hardboard and used for tool storage. A second pair of doors installed on the sidewall provides access to a separate and larger 10-ft. by 12-ft. storage area.

of siding to fit across the header at the top of the opening. Then you can resume installing full-length siding boards, making sure you stop and check for plumb every three or four boards. Rip the last board flush with the end stud and face-nail it to the wall frame.

Raise the wall

You'll need at least three people to help lift these walls off the floor deck. Be sure to keep children and pets well away from the site during this construction phase.

Slide the completed sidewall into position, with its bottom plate about 3 in. from the edge of the floor deck. Now you and your helpers must grasp the wall by the 2×4 top plate and slowly lift up the wall. When it's vertical, grab a sledgehammer and move to the outside. Standing at one end

of the wall, use the sledgehammer to tap in the very bottom of the wall to ensure it's up against the floor frame. Hold the sledgehammer against the siding while a helper on the inside drives a 3-in. screw down through the bottom plate and into the floor.

Now move to the center of the wall and repeat the process. Tap in the wall with the sledgehammer so the siding is tight against the floor framing, then hold it in position while it's screwed to the floor. Fasten the remaining end of the wall in a similar manner.

Drive two screws, equally spaced, between each pair of vertical blocking. Angle the screws slightly so they'll go through the wall plate and plywood floor deck and into the 2×6 perimeter joist below. Install a temporary 2×4 diagonal brace on each end of the wall to keep it from falling.

Lift the preassembled sidewall and stand it along the left edge of the floor. Keep a firm grip on the wall until it has been secured with screws.

Use a sledgehammer to push in the wall, then drive a 3-in. screw through the wall plate and into the floor frame below.

Build the remaining walls

With the first wall in place and temporarily braced with 2×4s to keep it from falling, the next step is to build the frame for the opposite sidewall. It's the same size as the first wall—82¼ in. high by 15 ft. 8 in. long—but it's much easier to build because it doesn't have any windows. Build it like the first wall, remembering to add the 3½-in.-wide plywood spacers to the end studs; select two perfectly straight 2×4s for the girts. Also, be sure to cut the siding so it overhangs the bottom wall plate by 7 in. Raise the second sidewall just as you did with the first one, secure it to the floor with 3-in. screws, and temporarily brace it in place.

Now build the rear wall to 9 ft. 5 in. wide. Unlike the sidewalls, this wall has no vertical blocking, but it does have two pieces of diagonal metal wind bracing. It's made up of only six 2×4s: a top plate, a bottom plate, two end studs, and two horizontal girts. Begin by cutting the 2×4 top and bottom wall plates to 9 ft. 5 in. long and the two end studs to 79¼ in. long. Attach the wall plates to the end studs with 16d nails, then check the wall frame for square by comparing the diagonals.

Assemble the opposite sidewall on the floor deck, then lift it up. Have one person outside the wall pushing in at the very bottom.

Cut two 2×4 girts to 9 ft. 2 in. and set them between the end studs (there's no need to add plywood spacers to the rear or front walls). Nail the first girt 21¼ in. up from the bottom wall plate, driving 16d nails through the end studs and into the girts. Then measure 21¼ in. from the first girt and nail the second girt in place. Use a portable circular saw to let in two pieces of diagonal metal wind bracing.

Nail on the siding just as you did for the previous walls. Set it flush with the top plate and allow it to overhang the bottom plate by 7 in. Erect the rear wall, which should fit exactly between the two sidewalls. Screw the bottom plate to the floor deck, as you did with the two sidewalls, then use a 4-ft. level to check both walls in each corner for plumb. Don't worry if the two walls aren't exactly flush; the corner boards will hide any offset. It's more important to make sure the walls are perfectly plumb. Secure each shed corner by screwing through the end

PRO TIP

Fasten the roof trusses to the walls with screws, not nails. That way, you can easily back out the screws if you need to reposition a truss or straighten a wall.

TRADE SECRET

Setting a gable-end truss in place takes teamwork—and a little help from below. Position a long 2×4 close at hand so that once a truss is rotated into place, the 2×4 can be used to slowly push the truss upright. This allows the workers on the ladders to concentrate on holding the truss on top of the end wall.

TRADE SECRET

The trusses are spaced 24 in. apart, but you don't need a tape measure to determine where they go; simply set each truss directly over a piece of vertical blocking. This helps transfer the weight of the roof to the wall frame and down through the floor frame to the skid foundation.

stud of the rear wall and into the end stud and blocking of the sidewall, using 3-in. decking screws spaced about 12 in. apart.

The front wall is built to the same dimensions as the rear wall: 82¼ in. high by 9 ft. 5 in. wide. Note that this wall has a 58-in.-wide by 72-in.-high rough opening for the door. Cut the top and bottom wall plates to 9 ft. 5 in. long and nail them to the 79¼-in. end studs. Don't worry about the rough opening for now; allow the bottom plate to span the entire length of the wall. It will be cut from the door's threshold after the wall is raised and screwed in place.

Frame the door's rough opening in the center of the wall frame. Establish the top of the rough door opening by installing a 61-in.-long header cut from a 2×4. Position the underside of the header 72 in. up from the bottom of the bottom wall plate. Install a 79¼-in.-tall king stud and a 70½-in.-tall jack stud on each side of the opening. The shorter jacks fit underneath the rough header that spans the top of the doorway. Secure each end of the header with two 3-in. screws driven through the king studs.

On each side of the doorway is a 27½-in.-wide section of wall. In each section, install two rows of horizontal girts, spaced 21¼ in. apart, just as you did for the other three walls. When the last wall has been framed and squared, it can be covered with V-jointed cedar siding. However, don't install any siding over the rough doorway opening or over the narrow space directly above the doorway. This 7¼-in.-tall opening is where the transom window will go.

After nailing on the siding, lift up the wall between the two sidewalls, tap it into position with the siding held tightly against the floor framing, and screw its bottom plate to the floor deck. Check the corners for plumb, then screw through the end studs of the front wall and into the end studs and blocking of the sidewalls.

Roof Framing

Framing a roof in the traditional manner—with individual rafters, collar ties, and ridge board—is a tedious, time-consuming chore that could easily take all day to complete. However, because you've already assembled the roof trusses, it should take about only 90 minutes to frame the entire roof.

You'll need at least three people for this job: one on the ground and two on top of the sidewalls. The person on the ground is responsible for handing up the trusses and for screwing the trusses to the top wall plate. The two people on top of the sidewalls set the trusses in place and hold them while they're screwed from below.

Start by installing the two gable-end trusses—the ones with the siding attached—to the front and rear walls. Then attach the seven standard trusses to the sidewalls. After the roof is framed, you can nail down the plywood sheathing, then install the shingles.

Set the gable-end trusses

Although standard roof trusses are much lighter and easier to install than gable-end trusses, don't make the mistake of installing them first. If you do, there won't be room to flip the gable-end trusses into place. You'll then have to lift up those heavy trusses from the outside, which is virtually impossible to do without the aid of a small army.

+ SAFETY FIRST

Climbing stepladders has become rather routine for most homeowners, but nonchalance can breed disaster. When using a stepladder to climb atop shed walls, make sure it's standing on a flat, dry surface that's free of debris and construction-site clutter. The top of the ladder should lean against a strong, flat surface, not balance on a wall stud.

Stand up the gable-end truss, then drive 3-in. decking screws up through the top wall plate and into the shoe plate on the truss.

Begin by carrying one of the gable-end trusses through the doorway and setting it upside-down on top of the sidewalls, with its peak pointing toward the floor. Slide the truss against the rear wall and check to make sure its exterior surface (the one sided in cedar) is facing in toward the center of the shed. With two helpers standing on ladders situated in opposite corners and close to the rear wall, swing the peak of the truss upward until the

helpers on the ladders can reach it. Then stand back while they slowly push the truss to its full, upright position.

With your helpers holding the truss upright, align the 2×4 shoe plate screwed to the bottom of the truss so it's flush with the top wall plate. Secure the truss by driving 3-in. decking screws up through the wall plate and into the shoe; drive one screw every 12 in. Carry in the second gable-end truss and install it over the front wall, using the same procedure.

Install the standard roof trusses

Standard roof trusses are much lighter and easier to install than gable-end trusses are, but you'll still need three people to lift them into position. Set up two ladders inside the shed so two people can work on opposite sidewalls.

✔ According to Code

In some towns, you'll be required to cover ply-wood roof sheathing with felt underlayment. Here, I nailed the roof shingles directly to the ply-wood because felt wasn't required or necessary; the shed is unheated and has a steeply pitched roof. Felt underlayment must be installed on any outbuilding with a roof slope of 4-in-12 or less.

TRADE SECRET

A Japanese-style handsaw is a good choice for cutting off excess cedar trim along the rake. This tool has a super-sharp blade that reduces splintering, which is important because you have to make the cut from the back. Unlike Western-style saws, Japanese saws—sometimes called SharkSaws®—cut on the pull stroke, and their ultra-thin blades produce very smooth cuts.

WHAT CAN GO WRONG

When the roof trusses are spaced 24 in. on center, there's a chance that the plywood roof sheathing will dip down or bow up along the unsupported edges between the trusses, resulting in a rather unattractive, wavy roof. One way to prevent plywood deflection is by using plywood clips, which are small H-shaped metal brackets. Install one clip on the plywood seam between each pair of trusses.

After erecting both gable-end trusses, install the seven roof trusses. Place one truss directly over each row of vertical blocking.

From outside the shed, hand up the first truss to your helper, who must then pull it onto the wall and slide it across to the helper on the opposite wall. Set the truss directly over the first line of vertical blocking. Working on the left-hand sidewall (the one with the windows), align the very end of the bottom chord on the truss with the outside edge of the wall's top plate.

With each helper standing on the bottom chord to keep the truss from shifting out of position, drive two 3-in. screws up through the top plate and into the bottom of the truss. Because the blocking is directly beneath the truss, you'll have

to drive the screws up at a slight angle, placing one screw on each side of the blocking. Don't fasten the other end of the truss just yet.

Continue installing trusses in this manner, making sure you place one over every line of vertical blocking. Remember: Fasten the trusses to the left-hand sidewall only, aligning the end of each bottom chord with the outer edge of the top plate.

After all seven trusses are in position and fastened on one side, you can begin fastening them to the opposite sidewall. This time, start with the truss in the middle of the wall. Using the bottom chord as a guide, push the wall in or out to align the end of the chord with the top wall plate. When the truss and wall are properly aligned, secure the truss with two 3-in. screws driven up through the plate and into the truss. Now attach the remaining six trusses, making sure each bottom chord lines up properly with the top plate.

When the trusses are in place, measure the length across the rafter tails from one gable-end truss to the other, and cut two long 2×4s to span

Sheathe the roof frame with ½-in. exterior-grade plywood. Fasten the plywood to the rafters with 1½-in. nails spaced 10 in. apart.

the distance. Nail these 2×4s—called subfascia—to the ends of the rafter tails with 3½-in. (16d) nails.

Install the roof sheathing

You'll need nine sheets of ½-in. thick exterior-grade plywood, either BCX or CDX grade, to cover the roof frame. Start with a full 4-ft. by 8-ft. sheet and lay it along the bottom edge of the roof. Align its long, horizontal edge with the 2×4 subfascia nailed to the rafter tails; place its vertical end in the middle of the rafter that forms the gable-end truss. Fasten the plywood along its lower edge with 1½-in. (4d) galvanized nails. Drive one nail through each rafter, making sure the edge of the plywood remains flush with the subfascia. If necessary, push the sidewall in or out to align the fascia with the plywood. What you're doing is using the plywood to square up the shed roof. Set another full sheet of plywood end to end with the first one and nail it flush with the subfascia.

Use a 4-ft. level to make sure each gable-end truss is plumb. Nail the vertical ends of the plywood to the rafters of each gable-end truss. Check

Cover the gable overhang at each end of the roof with a narrow strip of plywood. Cut the piece so it fits from the eave to the peak.

to make sure the rafters are 24 in. on center, then finish nailing the plywood sheets to the rafters with 1½-in. (4d) galvanized nails spaced about 10 in. apart.

You'll need to rip an 8¾-in.-wide strip of plywood to cover the gable overhang at each gable-end truss. Slide the plywood strip into place, making sure it fits snugly between the raised lip of

TRADE SECRET

Instead of architectural-style shingles, three-tab shingles are used for the starter course because they're only a single layer thick, so they lay perfectly flat. Each architectural-style shingle is laminated from two strips of roofing; its surface is too thick and uneven to use as a starter course. Because starter-course shingles are completely covered by the next course, you can use three-tabs of any color; they don't have to match the architectural-style shingles. Ask at the lumberyard for a bundle of shingles that has been damaged, ripped open, or discontinued. You should be able to buy it for practically nothing.

IN DETAIL

Most roof-shingle manufacturers recommend nailing each shingle 1 in. from each end and then 12 in. in between. Running lengthwise across each shingle is a thin stripe, called a fastening line. The roofing nails must be driven through this line to provide maximum hold-down strength and ensure that the large nail heads will be covered by the next shingle course.

After installing the 5-in.-wide cedar fascia board, nail on a 1×2 cedar fascia trim. Hold the trim flush with the top edge of the fascia.

the cedar rake board and the plywood sheet nailed to the rafters.

Continue to sheathe the rafters on both sides of the roof with plywood. When you get to the top, cut the plywood 1½ in. short of the peak on each side of the roof. That narrow opening will eventually be covered by a ridge vent, which will allow hot, stale air to escape from the shed.

Rip a 5-in.-wide fascia board from a cedar 1×6 and nail it to the subfascia with 2½-in.-long ring-shank siding nails. The upper edge of the fascia must be perfectly flush with the top surface of the plywood sheathing. Use a handsaw to trim the rake board on the gable overhang flush with the fascia. Then nail a 1×2 cedar fascia trim to the fascia. Again, keep the top edges flush. Cut the overhanging end of the 1×2 rake trim even with the fascia trim.

Roofing

The roof of this Colonial-Style Garden Shed is covered with architectural-style asphalt shingles, also called laminated shingles. The homeowners chose a brownish gray color that looks a bit like weathered wood shakes. Each shingle is 12 in. wide by 36 in. long.

This roof requires three squares (300 sq. ft.) of shingles. Four bundles of architectural-style

shingles usually cover one square, so buy 12 bundles. If you use standard three-tab asphalt shingles, you'll need only nine bundles, because three bundles of three-tabs typically cover 100 sq. ft.

For the starter course that covers the ends and edges of the roof, you'll need approximately 70 linear ft. of standard three-tabs. You'll also need about 18 linear ft. of hip-and-ridge shingles, which are nailed over the ridge vent that runs along the roof peak. Each bundle of hip-and-ridge shingles usually covers 20 linear ft.

Because each roof plane measures nearly 8 ft. from the drip edge to the peak, you'll have to install three metal roof brackets and a long 2×6 plank to reach the upper portion of the roof. Roof brackets are available at home centers and tool rental dealers.

Install the starter course

Before nailing down the architectural-style shingles, install a starter course of standard three-tab shingles along the sloped ends and lower edges of the roof. This initial course serves two functions: It adds an extra layer of protection to the perimeter of the roof and completely covers the plywood sheathing, so the sheathing won't show through the seams between the architectural-style shingles of the first course.

Start by snapping chalklines 11½ in. from the four ends and two edges of the roof. Set the first three-tab shingle on one of the chalklines that runs from the eave to the peak. Make sure the shingle is turned backward, with the tabs facing the center of the roof. That way, there will be solid coverage along the entire perimeter of the roof. Using 1¼-in. roofing nails, begin nailing the starter shingles along the chalkline, securing each shingle with three nails driven through the asphalt sealing strip that runs lengthwise down the center of the shingle. When installed, the 12-in.-wide starter shingles will extend past the roof deck by ½ in., creating a drip-edge overhang. At the peak, trim the shingle flush with the top edge of the plywood; don't cover the ridge-vent space.

Repeat this process to install the starter course along the three remaining roof ends. Then nail starter-course shingles along the bottom edge of the roof on both sides of the shed. Again, set the shingles right on the chalkline, but this time turn them upside-down so the tabs face the peak. Trim the last shingle to fit tightly against the shingle nailed to the end of the roof; don't overlap them.

Shingle the roof

With the starter course complete, begin installing architectural-style shingles, starting at the eave and working your way up toward the peak. Each

Use three-tab shingles for the starter course. Note that they're turned backward, with the tabs facing the roof's center.

Laying Shingles

There are two techniques for laying roof shingles. Most pros use the stair-step method, which allows them to install shingles across and up the roof at the same time. The advantage of this method is that you don't have to walk back and forth across the entire roof for each course; instead, you stand in one area. When one section is done, you move down several feet and start the next section.

The second method is more straightforward and, therefore, preferred by most do-it-yourself builders. Called in-line or straight-row shingling, this method consists of completing each course in its entirety from one end of the roof to the other. After each course, you go back to the beginning and start the next one. This method requires a lot more walking back and forth, but fewer mistakes are likely to occur.

Install the shingles with a 6-in. offset and a 5-in. exposure to the weather. Secure each shingle with four 1¼-in. roofing nails.

TRADE SECRET

You must leave a 1½-in.-wide opening between the plywood roof sheathing and the roof peak for a ridge vent to operate properly. However, if the roof is conventionally framed with regular rafters (not trusses) and a 2× ridge board along the peak, make sure the ridge opening is about 2½ in. wide.

IN DETAIL

Several types of ridge vents are available, but I prefer the flexible polyester-fiber type, which is easier to handle, cut, and install than the rigid aluminum and plastic vents are. Plus, it does a better job of keeping out insects and wind-driven rain. Flexible ridge vents are sold at most home centers and lumberyards in 20-ft. and 50-ft. rolls.

Lay the first course of architectural-style shingles directly over the starter course. Be sure to overlap all vertical seams by at least 6 in.

shingle must be secured with four 1¼-in. roofing nails. Read the installation instructions printed on the shingle-bundle wrapper for specific information about where to place the nails.

Starting at one end of the roof, begin nailing down the first course of shingles. Set each shingle directly on top of the starter course, making sure it's flush along the drip edge. Overlap all vertical seams in the starter course by at least 6 in. Continue laying full-size shingles until you reach the end of the roof, then cut the shingles to length, allowing for the ½-in. overhang at the end.

Next, cut 6 in. off a shingle and use it to start the second course. Trimming the first shingle to 30 in. long automatically staggers the seams between the first and second course by 6 in. Set the first shingle in the second course flush with the end of the roof and align its bottom edge to create a 5-in. exposure to the weather. Again, check the package label for specific information regarding shingle alignment and exposure. Your shingles may be different from the ones installed here. Continue the course by nailing down full-size shingles.

Cut 12 in. off another shingle and start the third course with it. As before, trimming the shin-

gle to 24 in. long will create a 6-in. offset with the seams in the second course. Repeat this pattern as you make your way up the roof. Start the fourth course with an 18-in.-long shingle, the fifth course with a 12-in.-long shingle and the sixth course with a 6-in.-long shingle. When you reach the seventh course, start with a full-size shingle and begin the pattern all over again.

If you maintain a consistent 5-in. exposure to the weather, the last course of shingles will overlap the peak by about 6 in. Use a utility knife to trim the shingles flush with the edge of the plywood roof sheathing.

Add the ridge vent

A ridge vent is installed over the narrow space that runs along the roof peak. This simple item is an important part of the shed's ventilation system. Fresh air flows up through the soffit vents—which will be installed under the eaves—and forces air trapped in the shed to exit from the ridge vent.

Begin by trimming the upper course of shingles flush with the plywood sheathing on both sides of the peak. Next, unroll the

Use a sharp utility knife to carefully trim the shingles flush with the plywood sheathing on each side of the roof peak.

Roof Brackets

After you've shingled halfway up the roof, stop and install three roof brackets. Nail each bracket to a rafter, not just to the plywood.

Lay a sound 2×8 plank across the three roof brackets and continue nailing shingles all the way to the peak.

After you've shingled about halfway up the roof, you'll need to stop and install three metal roof brackets so you can safely shingle the upper part of the roof. Place one bracket about 12 in. from each end of the roof and set the third one in the middle. Fasten each bracket with two 16d nails driven through the sheathing and into a rafter. To ensure the brackets are arranged in a straight line, measure down from the top edge of the roof sheathing an equal amount for each bracket. Set the top edge of the brackets on the marks. Set a long, straight 2×8 into the brackets, making sure it fits under the upturned lip at the front of each bracket. Then secure the plank by driving a 1½-in. nail (4d) through the hole in the front of each bracket. Carefully climb onto the plank and continue nailing down shingles, working your way toward the peak.

When it comes time to remove the brackets, start by pulling out the nails that hold the plank in place. Lift the plank out of the brackets and hand it down to a helper on the ground. Next, strike the bottom end of each bracket with a hammer to slide it up the roof about ½ in. That will center each nail head over a large hole in a keyhole slot in the bracket. Pull up to free the brackets.

Gently lift up the shingle and hammer down the nail heads. If you can't lift the shingle high enough to reach the nails, don't force it, or you may crack the shingle. Instead, slide a flat prybar under the shingle and rest it on top of the nail head. Give the bar a couple of good whacks with the hammer to drive the nail home.

10-in.-wide vent along the center of the ridge, and fasten it to the roof with 2-in.-long roofing nails (or per the manufacturer's directions). Drive one nail through the vent and into each rafter. Tap the nails just far enough to hold the vent on the roof; don't compress or crush the fibers. Then, starting at one end and working toward the other, cover the vent with the hip-and-ridge shingles. Overlap the shingles to produce a 5-in. exposure to the weather and fasten each one with two 2-in. roofing nails. Again, be careful not to overdrive the nails and crush the vent.

The final step is to install a perforated aluminum soffit vent under the overhanging eave on each side of the shed. The vents are sold in 12-in.-wide sheets that must be cut lengthwise down the middle. Score the sheet a few times with a sharp

Unroll the 10-in.-wide ridge vent and fasten it with 2-in. roofing nails. Don't drive the nails too far in, or you'll crush the vent.

PRO TIP

To avoid measuring mistakes, bring the completed frame of the transom window to a glass shop and have the glazier custom-cut the glass to fit.

TRADE SECRET

To give the doorway a more finished look, cover the exposed rim joist beneath the threshold with two pieces of V-jointed 1×6 cedar siding. Cut the length to span the distance under the doorway, and rip the tongue off of one board to form a straight edge. Set the boards in place, making sure the straight-edged upper board is flush with the top of the plywood floor.

TRADE SECRET

When trimming a window, keep the following rules in mind:

- Cut the head casing and windowsill to the same length.
- Plane a slight bevel in the top surface of the sill so that rainwater drains away from the window sash.
- Cut the apron to equal the distance across the side casings, as measured from the outside edges.

Conceal the ridge vent with hip-and-ridge shingles. Center each shingle over the vent, then attach it with two 2-in. roofing nails.

Install a perforated aluminum soffit vent on the underside of the overhanging eave along each side-wall. Secure it with 1-in. screws.

utility knife, then bend it until it snaps into two 6-in.-wide pieces. Slip the vent under the eave, fitting it between the siding and the fascia board. Secure it to each rafter and to the 2×4 subfascia with 1-in. decking screws.

Windows and Exterior Trim

This shed has three windows: two barn sash windows on the left-hand sidewall and a transom window above the pair of swinging doors on the gable end. The transom is a fixed window, meaning it doesn't open. Both 2-ft. by 4-ft. barn sashes tilt in for ventilation.

Barn sash windows are sold at most lumber-yards and farm-supply outlets, but they're not usually kept in stock, so you'll have to order them. It may take anywhere from a few days to a couple of weeks for the windows to arrive. However, before you can install them, you must trim the exterior window openings with roughsawn cedar and attach an interior window stop. Each sash is held in place by the exterior trim and a 1×2 stop, which is fastened to the sill.

The transom window, which measures 9¾ in. tall by 59 in. long, must be built on site to fit the space above the doors. Fortunately, the window is very easy to make and consists of little more than a wooden frame and two pieces of double-strength glass.

Make the transom window

Cut the parts for the frame of the transom window from leftover 1×6 cedar siding. A table saw is the tool of choice for cutting the parts to size, but you can also use a portable circular saw.

Start by cutting off the tongue-and-groove edges from two 6-ft. lengths of ¾-in.-thick siding, then rip the boards to the proper width. Note in the drawing below that the seven vertical mullions and the bottom horizontal rails are 1½ in. wide, but the top rail is 2 in. wide. That's because when

the window is installed, the top rail will extend ½ in. behind the siding on the gable-end truss, resulting in a consistent 1½-in.-wide frame. Crosscut the mullions to 9¾ in. long and make the two rails 59 in. long.

Next, cut a ³/₁₆-in.-deep by ⅞-in.-wide rabbet along the rear edge of the two long rails. These rabbets create a recess for installing the glass panes. You can cut rabbets with a router, a radial-arm saw, or a rabbeting plane, but I prefer to use a table saw fitted with a dado blade.

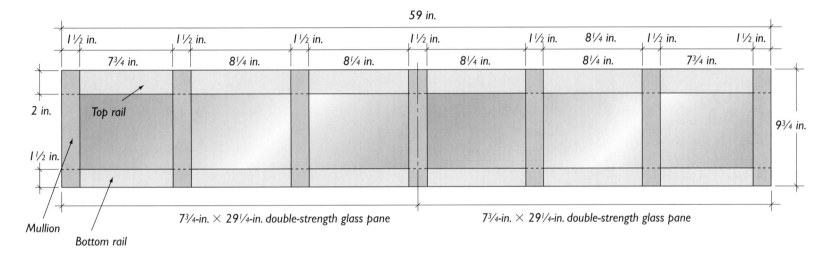

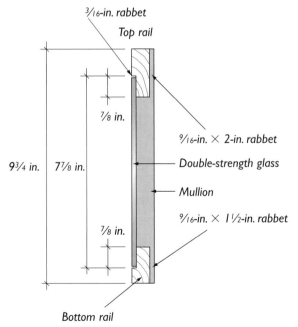

Transom Window

The narrow transom window installed above the batten doors consists of a wooden frame and two panes of glass. Note that all the frame parts are 1½ in. wide, except the top rail, which is 2 in. wide. Rabbets are milled in the rear of the frame to accept the glass panes.

PRO TIP

Paint the pine frames of the barn sash windows before installing them. Apply a coat of primer followed by two top coats of paint.

IN DETAIL

Barn sash windows are ideal for outbuildings because they're affordable, easy to install, and readily available. However, their pine frames must be refinished every few years, and they come in only one basic style. To avoid the drudgery of scraping and repainting, consider windows that are clad in aluminum or vinyl. Some manufacturers even offer clad windows in a variety of colors and shapes.

TRADE SECRET

A ramp is useful for rolling equipment in and out of the shed, even when the doorway is only a few inches above the ground. This ramp is made from a couple of pressure-treated 2×12s. The nice thing about this type of ramp is that you can store it inside the shed until the next time you need it.

Also, cut rabbets in both ends of each mullion. These rabbets allow the vertical mullions to fit around the top and bottom rails and lie flat against the glass. However, because the top rail is wider than the bottom rail, the rabbets are different sizes. In the upper end of each mullion, cut a $\frac{9}{16}$-in.-deep by 2-in.-wide rabbet. In the lower end, cut each rabbet $\frac{9}{16}$ in. deep by 1½ in. wide.

Now assemble the frame, using water-resistant carpenter's glue and ¾-in.-long wire nails. Begin by gluing and nailing the two end mullions to the outside surface of the top and bottom rails to form the rectangular frame of the window. Secure both ends of each mullion with two wire nails. Check the frame for square by comparing the diagonals. Then attach the five remaining mullions to the outside surface of the frame, installing the first one 7¾ in. from the end mullion; set the remaining mullions 8¼ in. apart, as shown in the drawing on p.161.

The frame accepts a glass pane that measures 7¾ in. by 58½ in. Although you can order a single piece of glass from a glass shop, I've found out the hard way that it's much easier to transport and install two shorter pieces of glass. Therefore, I suggest getting two pieces of double-strength glass, each measuring 7¾ in. by 29¼ in. To install the glass, place the frame face down on a flat surface and run a thin bead of clear silicone adhesive along the rabbets cut in the rails. Also, run a small bead of

silicone on each mullion. Carefully set the glass panes in the frame, then lightly press them into the adhesive. Allow the silicone to cure overnight.

To install the transom window, simply raise it above the doorway and slip its top rail behind the siding on the gable-end truss. Press the frame's bottom rail against the 2×4 rough header, then attach the window with ten No. 4, 1½-in. brass screws. Drive five screws, equally spaced, through the bottom rail and five more through the top rail. Be sure to predrill $\frac{3}{16}$-in.-dia. screw-shank clearance holes to avoid splitting the frame. There should be about ½ in. of the header visible below the window frame. That small reveal provides a flat surface for the doors to close against.

Trim the barn sash windows

The exterior of each 2-ft. by 4-ft. barn sash window is finished with five pieces of trim, all cut from roughsawn red cedar. The simple pine sash is available at most lumberyards in various square and rectangular sizes. Installed flat across the bottom of the opening is a windowsill cut from a 5/4 by 6-in. cedar board. On each side of the window is a 3-in.-wide vertical side casing cut from a 1×4. Above the opening is a head casing, which is made from a 1×6 and accented with a couple of 1×2 battens and a decorative keystone. Running horizontally beneath the sill is a 3-in.-wide apron cut from a 1×4.

Mount the transom window in the narrow opening above the doorway. Screw it to the 2×4 rough header and top wall plate.

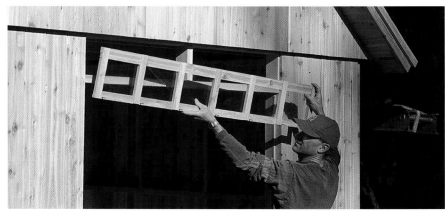

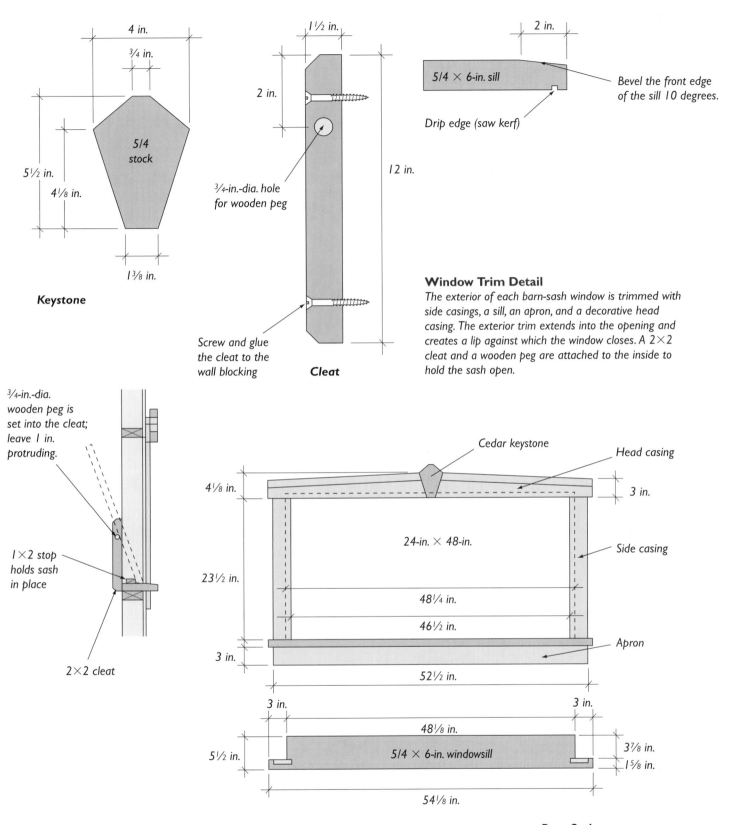

4 in.

3/4 in.

5/4
stock

5 1/2 in.

4 1/8 in.

1 3/8 in.

Keystone

1 1/2 in.

2 in.

12 in.

3/4-in.-dia. hole
for wooden peg

Screw and glue
the cleat to the
wall blocking

Cleat

2 in.

5/4 × 6-in. sill

Bevel the front edge
of the sill 10 degrees.

Drip edge (saw kerf)

Window Trim Detail
The exterior of each barn-sash window is trimmed with
side casings, a sill, an apron, and a decorative head
casing. The exterior trim extends into the opening and
creates a lip against which the window closes. A 2×2
cleat and a wooden peg are attached to the inside to
hold the sash open.

3/4-in.-dia.
wooden peg is
set into the cleat;
leave 1 in.
protruding.

1×2 stop
holds sash
in place

2×2 cleat

Cedar keystone

Head casing

4 1/8 in.

3 in.

24-in. × 48-in.

Side casing

23 1/2 in.

48 1/4 in.

46 1/2 in.

3 in.

Apron

52 1/2 in.

3 in.

3 in.

48 1/8 in.

5 1/2 in.

5/4 × 6-in. windowsill

3 7/8 in.

1 5/8 in.

54 1/8 in.

Barn Sash

Lift the built-up head casing over the window and set it on top of the vertical side casings; secure it with 1½-in. siding nails.

WHAT CAN GO WRONG

Over time, shed doors tend to sag and rub together, making it difficult to close them. This is often caused when hinge screws become loose or strip out their holes. To remedy this problem, start by removing the hinge-cup screws and shimming the door back into its original position. Reattach the hinge cups with longer screws. Drive the screws into solid wall framing, not just the siding.

TRADE SECRET

When searching for ways to utilize a shed's storage space, don't overlook the backs of the doors. These broad surfaces offer a convenient place to hang long-handled tools. A pivoting-hook storage rack, like the one shown above, works particularly well. It's designed for brooms and mops, but it'll hold rakes, hoes, and other garden tools. Just slip the tool's handle under the S-shaped hook.

Nail the 3-in.-wide cedar apron to the shed wall directly beneath the windowsill. Align the apron ends with the side casings.

It's important to note that the head and side casings aren't nailed flush with the edge of the window opening; instead, they extend ⅞ in. into the opening. That creates a lip along the top and sides of the opening to hold the sash in place. On the interior of the rough opening are two pieces of trim. Running lengthwise across the top of the sill is a horizontal window stop cut from a cedar 1×2. It holds the bottom of the sash in place.

Screwed to the wall blocking on each side of the opening is a 12-in.-long vertical 2×2 cleat. A short wooden peg projects from the cleats 1 in. into the opening. When the barn sash is tilted open, it comes to rest against the pegs. The sash itself isn't hinged or screwed in place at all. It simply sits on the windowsill between the

window stop and the exterior trim and is held shut with a barrel bolt attached to the interior surface of the sash.

Start by cutting the windowsill to 54⅛ in. long. Saw a 3-in.-wide by 3⅞-in.-deep notch in each end. Set the sill in place on top of the 2×4 rough sill and secure it with six 2½-in.-long siding nails. Next, cut the two vertical side casings to 23½ in. long. Set each one on top of the sill, making sure it overlaps the opening by ⅞ in. Nail the casings to the sidewall.

The head casing, similar to the ones found on old barns, comes to a peak in the center, which helps shed water. But rather than just using a single flat board, this casing is dressed up a bit with two 1×2 battens and a center keystone block cut from 5/4 cedar.

Start by cutting a 1×6 to 54⅛ in. long, then rip it to 4⅛ in. wide. To create the pediment shape, measure 3 in. up from each end and make a mark. Draw a line from the top center of the board to the marks on each end. Use a portable circular saw to make angled cuts from the center point to the 3-in. marks. The result is a board that's 3 in. wide at the ends and 4⅛ in. wide in the middle.

Make the decorative keystone from a 4-in. by 5½-in. block of 5/4 cedar. Using the drawing on p. 163 as a guide, mark the double-angled shape on the block. The safest way to execute the angled cuts

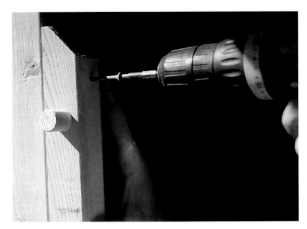

Mount a 2×2 cleat on each side of the window. The wooden peg protruding from the cleat supports the sash when it's tilted open.

After installing the window trim, slip the barn sash into place. Check to make sure it tilts in and securely latches closed.

is with a sabersaw or handsaw. The block is a bit too small to cut safely with a power miter saw.

Attach the keystone to the center of the head casing by driving three 1½-in. screws through the back of the casing. Install two 1×2 batten strips on the face of the casing. Cut each one to fit from the keystone to the end of the casing. Attach each batten with three 1½-in. siding nails; be sure to predrill ⅛-in.-dia. pilot holes, or the narrow battens will split. Set the head casing on top of the side casings and secure it with four equally spaced siding nails.

Make the apron by first cutting a 1×4 to 52½ in. long. Then rip it to 3 in. wide and nail it in place underneath the windowsill. Make sure that the ends of the apron are aligned with the outside edge of the side casings.

Next, cut a 1×2 to 48⅛ in. long to use as the window stop. To make sure you install the stop in the correct position, use the barn sash as a guide. Set the sash in the opening from the inside and hold it tightly against the lip formed by the exterior trim. Then set the window stop across the sill and press it against the sash. Draw a pencil line along the edge of the stop and onto the sill. Remove the sash, align the stop with the pencil line, and nail it to the sill with four 1½-in. siding nails.

Nail together two cedar boards to form a corner cap. Place the cap on the shed corner and secure it with 2½-in. siding nails.

BUILDING BATTEN DOORS

The traditional batten doors on this shed are made from the same V-jointed, tongue-and-groove cedar boards used for the siding. The boards in each door are held together by three horizontal battens and two diagonal battens, which are screwed across the back. The 4¼-in.-wide battens are also cut from the V-jointed cedar boards. This particular style is called a Z-batten door because the diagonal battens form a Z-shaped pattern. Each door measures 29¼ in. wide by 74 in. high. Here are the basic steps for building each one:

1. Cut six tongue-and-groove 1×6s to 74 in. long. Use a table saw to rip the tongue off one board and the groove off another.

2. Lay the boards on a workbench with their roughsawn surfaces facing down. Align the ends flush, then use pipe clamps to draw the boards together into a single panel. Make sure the first and last boards are square at the edges.

Position six V-jointed, tongue-and-groove cedar boards with the rough sides down, then clamp them together to make one door panel.

3. Cut three horizontal rails from a 4¼-in.-wide cedar board; make each one 27⅞ in. long. Apply construction adhesive to the backs of the rails, then attach them with 1¼-in. decking screws.

4. Cut a Z-batten to fit diagonally from the bottom rail to the center rail. Run a bead of adhesive across the back and set it in place. Cut a second Z-batten to fit between the center rail and the upper rail. Run this one in the opposite diagonal direction of the lower Z-batten.

Apply a zigzag of construction adhesive on the backs of the three horizontal rails. Cut the 4¼-in.-wide rails from cedar siding.

5. Screw a 2-in.-wide astragal strip to the latch side of the left-hand door. This overhangs the door edge by 1 in. to form a support strip for the right-hand door to close against.

6. Hang each door with a pair of 12-in. strap hinges. Bolt the strap portion to the door with carriage bolts, then set the door in the opening. Use shims to create the proper clearance at the top and bottom of the door.

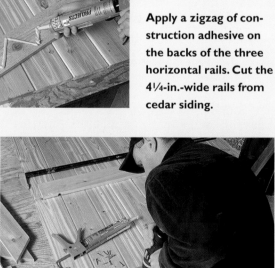

7. Slip a cast-iron hinge cup onto the top and bottom of the hinge pin protruding from each strap. Fasten the cups to the shed wall with the screws provided.

Fasten the rails with 1¼-in. decking screws. Be careful not to overdrive the screws, or they'll poke through the other side.

Install a second Z-batten in the opposite diagonal direction. Cut it to fit snugly between the top and center rails.

Mount a 2-in.-wide cedar astragal strip on the edge of the left-hand door. Allow the strip to overhang the door edge by 1 in.

This traditional-style strap hinge has a pin that fits into two cast-iron cups. Slip the cups onto the pin, then screw them in place.

The last window-trim pieces to install are the 12-in.-long cleats, which are cut from a 2×2. Bevel-cut the ends at a 45-degree angle, then measure 2 in. down from one end and bore a ¾-in.-dia. by 1-in.-deep hole in the side of each cleat. Cut off two 2-in.-long pegs from a length of ¾-in.-dia. hardwood dowel rod. Glue the pegs in the holes. Hold each cleat against the vertical 2×4 blocking on each side of the window opening. The pegs must face the opening and the bottom ends of the cleats must be flush with the top of the 2×4 rough sill. Secure each cleat with two 3-in. screws.

Next, mount a 2-in. sliding barrel bolt on the top center of the sash. Slip the sash into place from the inside. Set it on the sill, with its lower edge between the exterior trim and the window stop. Finally, bore a ⁵⁄₁₆-in.-dia. hole for the barrel bolt in the 2×4 rough header above the sash.

Install the corner boards

The cedar corner boards that cover the four corners of the shed are the last pieces of exterior trim to be installed. You can cut and install these eight boards one at a time, but it's much easier to pre-assemble a corner cap for each shed corner by first nailing together two boards.

Each 90-in.-long corner cap is made from a roughsawn cedar 1×4 and 1×6. Rip the 1×6 to 4¼ in. wide, then nail it to the narrower 1×4 with 2½-in.-long siding nails spaced 12 in. apart. This creates an L-shaped corner cap measuring 4¼ in. square. Slip the corner cap into place, making sure it fits tightly against the soffit vent on the sidewalls and under the siding on the gable-end truss on the front and rear walls. The bottom end of the corner boards should be flush with the siding. Secure the corner cap to the shed with 2½-in.-long siding nails spaced 12 in. apart.

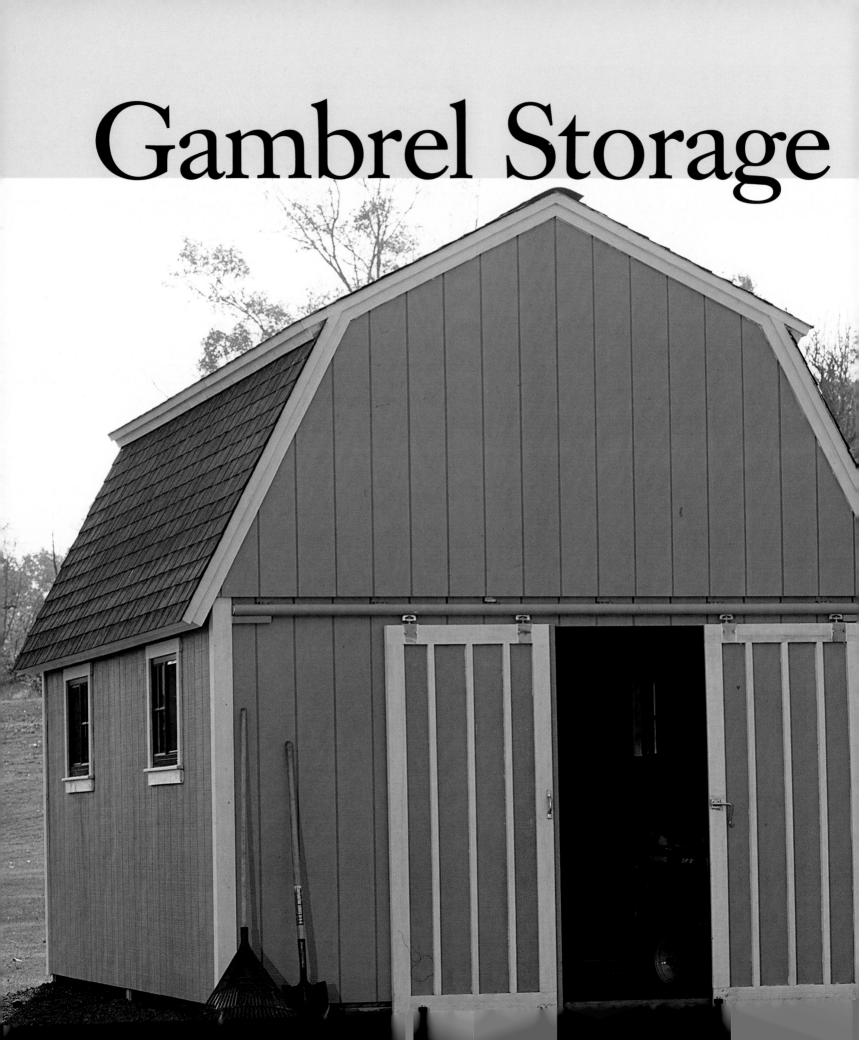

Gambrel Storage

CHAPTER SEVEN

Barn

Traditional barn architecture and modern building methods come together in this spacious 12-ft. by 20-ft. storage barn, which features a gambrel-style roof, grooved plywood siding, aluminum windows, and pair of sliding doors. At 240 sq. ft., the building is supported by a pole-barn foundation consisting of six 4×4 treated posts set on concrete footings. Note that the entire structure is suspended from the posts; there's no wooden floor.

The ground inside the barn is covered with processed stone, creating a floor that's just about even with the ground outside. That makes it easy to store lawn tractors, farm machinery, and trailers. Another benefit is that the gambrel roof creates a spacious storage loft above the ceiling joists. Need one more reason to build this barn? Consider this: It'll make the neighbors really jealous. (To order plans for the Gambrel Storage Barn, see Resources on p. 198.)

TRADE SECRET

Gas-powered sod cutters are typically rented by the day, even if you use it for only an hour. That's not very cost-effective, but here's one possible solution: Reserve the machine two weeks in advance and ask the rental dealer for an hourly or half-day rate. If the dealer can rent out the machine for the other portion of the same day, he'll be more likely to give you a better rate.

IN DETAIL

The unusual shape of a gambrel roof was named for an equally unusual object: the hind leg of a horse. In 17th-century England, someone thought that the double-sloping angle of the roof resembled the bent portion of a horse's leg, which was called a hock or gambrel. From that moment on, it was known as a gambrel roof.

Use a gas-powered sod cutter to quickly remove sod from the building site. Roll up the sod strips and transplant them elsewhere.

Site Prep and Footings

The local building code required a permanent, frost-proof foundation for this storage barn because it's more than 200 sq. ft. Codes differ from region to region, but chances are that you, too, will need to build a frost-proof foundation. If you're planning to use the barn as a woodshop or carriage house, consider pouring a monolithic concrete-slab foundation. For general storage, the pole-barn foundation used on this barn works fine—and it's a whole lot easier to build.

Remove the sod

Before you can start laying out the foundation, you must remove the sod from the building site. That not only creates a more stable surface for the

processed-stone floor, but it also gives you the opportunity to rake the ground flat and somewhat level. Note that it's necessary to remove the sod from an area slightly larger than the footprint of the building; in this case, you must clear a 14-ft. by 22-ft. area. That will create a well-draining, grass-free buffer zone around the perimeter of the barn.

There are two basic ways to remove sod: by hand with a flat-blade shovel or with a gas-powered sod cutter. The choice is easy, considering that you must clear an area that's more than 300 sq. ft.: Go out and rent a sod cutter. This self-propelled machine slices through sod like a hot knife through butter. It should take about an hour to remove all the sod.

Mark out the 14-ft. by 22-ft. area using spray paint or white flour sprinkled from a can. Then run the sod cutter back and forth from one end of the building site to the other. Have a helper roll up and remove the sod strips.

Set up the batter boards

Once the sod is removed, the next step is to determine exactly where to dig the six footing holes. One of the easiest and most accurate methods for laying out footing holes is to use strings and batter boards. These pairs of vertical stakes are placed a few feet beyond each corner of the foun-

Gambrel Storage Barn

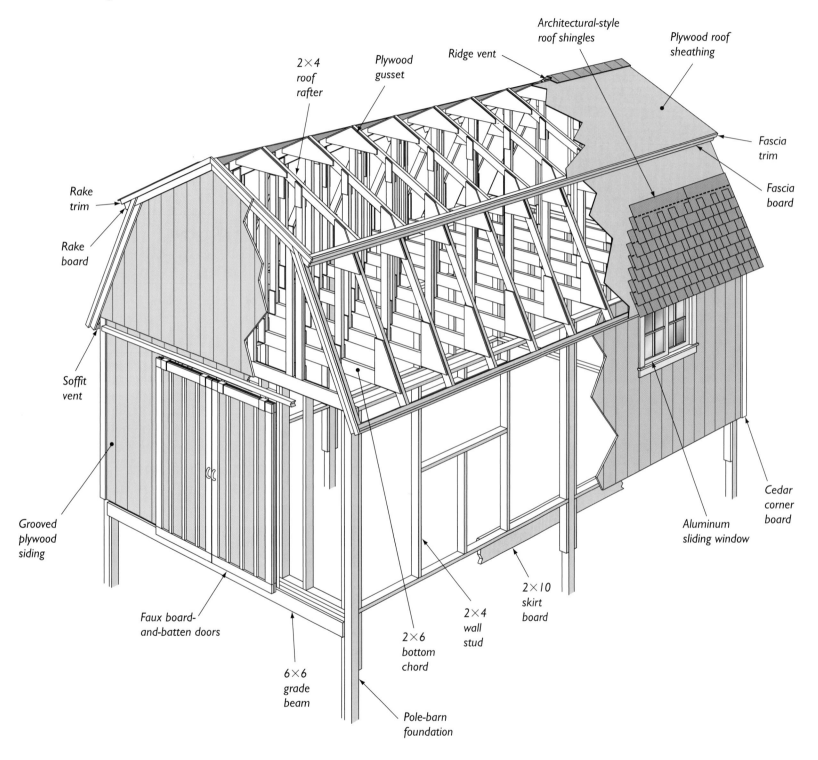

Architectural-style
roof shingles

Plywood roof
sheathing

Ridge vent

2×4
roof
rafter

Plywood
gusset

Fascia
trim

Fascia
board

Rake
trim

Rake
board

Soffit
vent

Cedar
corner
board

Grooved
plywood
siding

Aluminum
sliding window

Faux board-
and-batten doors

2×10
skirt
board

2×4
wall
stud

2×6
bottom
chord

6×6
grade
beam

Pole-barn
foundation

Complete the batter board by screwing a 1×4 crosspiece to the vertical stakes. Use a 2-ft. level to ensure that the crosspiece is level.

PRO TIP

Use nylon mason's line, not cotton string, to lay out the batter boards. The nylon line is very strong, stretches taut, and doesn't twist and tangle when rewound.

TRADE SECRET

A can of spray paint provides the easiest, most accurate way to mark the ground for the footing holes. Hold the fiber-form tube in place, then spray a coat of paint around the base of the tube. Move the tube out of the way to reveal a clearly marked circle. You can use regular spray paint, but for better results, buy a can equipped with a tip specially designed to spray upside-down.

IN DETAIL

Centuries ago, batter boards were used by stonemasons to check the receding upward incline—or "batter"—of a stone wall or bank. Today, batter boards provide pro builders and do-it-yourselfers with a quick and highly accurate way to lay out concrete footings, foundation walls, and deck posts.

dation, then strings are stretched tightly between them to represent the rectangular outline of the foundation. Key dimensions are then taken off the strings with a tape measure and plumb bob.

Using the drawing below as a guide, install the first batter board near one corner of the foundation. Drive in the two vertical stakes, spacing them about 3 ft. apart, then screw a crosspiece to the stakes, making sure it's level. Tack a small nail in the top edge of the crosspiece centered between the stakes; leave about 1 in. of the nail protruding. Install the opposite batter board in a similar manner, but don't attach the crosspiece

just yet. Tie a string to the nail on the first batter board and pull it across to the second one. Hang a line level on the string and raise or lower the crosspiece of the second batter board until the string is level. Screw the crosspiece to the stakes and drive a nail in the top edge. Pull the string taut and tie it off. Repeat this procedure to install the remaining three pairs of batter boards.

After stretching all four strings taut between the batter boards, check them for square by holding up a framing square where the strings intersect at 90-degree angles. The corner is square when the strings align with the edges of the fram-

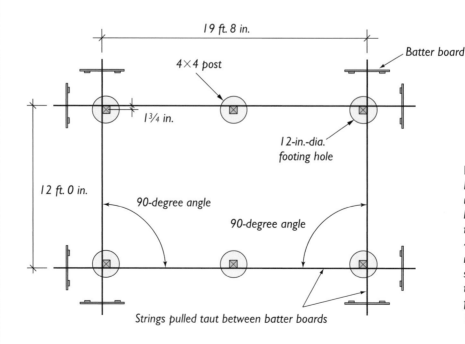

Foundation Plan
Batter boards and strings are used to locate the six footing holes. The strings represent the outside surface of each 4×4 post. Therefore, you must measure in from the strings 1¾ in. (half the thickness of the 4×4s) to find the center of the 12-in.-dia. footing holes.

Hook a line level onto the string, then raise or lower the crosspiece until it's level with the crosspiece on the opposite batter board.

To find the center of the 12-in.-dia. footings, measure 1¾ in. in from the layout line and drop a plumb bob to mark the ground.

ing square. If necessary, untie a string and move it left or right along the crosspiece. When the corner is square, reattach the string. Repeat this procedure for the other three corners. Now, check for square once again, but this time measure the diagonals of the rectangle formed by the four strings (you'll need a helper for this). When the two measurements are equal, the layout strings are square.

Note from the drawing on the facing page that the 144-in.-wide by 236-in.-long rectangle formed by the strings represents the outer edge of the 4×4 posts. To find the center of the footing holes (and the posts), measure in from the string 1¾ in., which is half the width of the 3½-in.-sq. posts. Hold the end of the tape measure against the string, making sure you don't deflect the string. Line up the plumb-bob line with the 1¾-in. graduation on the tape measure and drop the bob to mark the ground. The mark represents the center of the 12-in.-dia. footing hole and 4×4 post. Use this same technique to mark the locations of the remaining five holes.

Dig and pour the concrete footings

Each of the six footing holes must be lined with a 12-in.-dia. fiber-form tube. The tubes come in 12-ft.-long sections and are sold at most lumber-

yards and home centers. Use a handsaw or a sabersaw to cut the tube into pieces 6 in. longer than the depth of the hole. In this case, I had to dig 42 in. to reach the frost line, so I cut the tube into 48-in.-long pieces.

Stand a fiber-form tube over one of the footing-hole center marks made by the plumb bob. Peer down inside the tube and center it directly over the mark, then score the ground around the perimeter of the tube with a shovel or mark it with spray paint. Dig the hole to the desired depth, using both a shovel and a post-hole digger. Lower the fiber-form tube into the hole and backfill around it. As you're backfilling, have a

---

✓ According to Code

When building the foundation, be sure to dig the footing holes down to the frost line, but don't pour the concrete until after the building inspector has had a chance to inspect the holes. That's the building department's only assurance that the holes are dug to the proper diameter and depth.

PRO **TIP**

If you can't dig the footing holes with a shovel or post-hole digger in ground that's too hard or rocky, an excavator can bore the holes with a power auger.

TRADE SECRET

The tops of all the 2×4 cleats must be at the same height: 9 in. above the finished grade. The easiest way to do this is to cut the cleats about 12 in. too long, then attach them to the posts without nails in the top 12 in. or so. After the posts are in place, stretch a string across them and use a line level to level the string. Mark the cleats where indicated. Lift out the posts and cut the cleats at the marks.

Dig the footing holes down to the frost line (42 in., in this case). Cut the fiber-form tube to length and drop it into the hole.

Backfill around the outside of the fiber-form tube. Use a 2-ft. level to make sure the tube remains plumb.

+ SAFETY FIRST

Wet concrete looks harmless enough, but it's actually quite caustic. It can irritate skin and cause painful rashes. When mixing and pouring concrete, always wear eye protection, gloves, long sleeves, and pants (not shorts). If any concrete splashes onto bare skin, immediately wash it off with soap and water, then apply moisturizer.

helper hold a 2-ft. level against the inside of the tube to make sure it remains plumb. Dig the remaining five footing holes and install the fiber-form tubes.

Next, dump a bag of ready-mix concrete into a wheelbarrow, add water according to the directions, and blend it smooth with a hoe. Pour the entire contents of the wheelbarrow into one of the tubes to form a footing at the bottom of the hole. Be careful not to hit the protruding tube with the lip of the wheelbarrow or shovel, or you may knock it out of position. Pour the rest of the concrete footings, using one bag of concrete per hole. Allow the concrete to cure for at least 24 hours before setting the posts.

Wall Framing

The wall system of this pole barn consists of four 4×4 corner posts, two 4×4 intermediate posts, four sidewall sections, and two gable-end walls. The six posts are set on top of the concrete footings. The wall frames are assembled from 2×4s and plywood siding, then set between the posts. Two sidewall sections are needed to form each side of the 20-ft.-long barn. Also, there are a total of five window openings: one in each sidewall section and another in the rear gable-end wall.

Install the posts

Each of the six posts is made from a 12-ft.-long pressure-treated 4×4. Nailed to two sides of each post are shorter 2×4 pressure-treated cleats. Later, the walls will be set between the posts and on top of the cleats.

Start by making the two intermediate posts, which will be installed in the middle of the sidewalls. Cut 2×4 cleats to extend from the bottom end of the posts to about 12 in. above the finished grade, which will be the ground level at the doorway when the barn is completed. (The

Mix up a bag of concrete in a wheelbarrow and dump it into the hole to form a footing. Use one bag of concrete per hole.

Use 3½-in.-long galvanized nails to attach 2×4 cleats to the 4×4 posts. Nail the cleats to opposite sides of the two intermediate posts.

✔ According to Code

A permanent, frost-proof foundation—such as this pole-barn foundation—must satisfy more building codes than an on-grade type. Among other things, the inspector must approve the diameter, depth, and locations of the footing holes; the size of the concrete footing; the method used to secure the posts in the holes; and the way the walls are attached to the posts.

cleats will later be marked for trimming to length.) Using 3½-in. (16d) galvanized nails, attach the 2×4 cleats to opposite sides of each intermediate post, but don't nail within 12 in. of the top of the cleat (that part will be trimmed off later). Set the intermediate posts in the footing holes in the middle of the sidewalls.

Set the Corner Posts

Next, nail two 2×4 cleats to each corner post, but this time fasten them to adjacent sides, not opposite sides, of the post. This is necessary because the walls meet the corner posts at a 90-degree angle. Set the corner posts in the footing holes, again making sure the cleats extend from the bottom end of the posts to about 12 in. above the finished grade. Temporarily screw a long 2×4 across the three posts along each side of the foundation to hold them in place, but don't backfill around the posts just yet.

Find the post situated at the highest corner on the site and make a mark on the cleat 9 in. above the finished grade. Using that mark as a reference point, stretch a mason's string to the post at the opposite corner. Use a line level to level the string, then mark the cleats at each post where the string crosses the cleats. Repeat this procedure to mark the cleats on the remaining posts. Lift the posts out of the holes and use a circular saw to trim the cleats to the lines.

You'll also need to add one additional cleat to the inside surfaces of each of the two corner posts on the gable-end wall that will be framed for the sliding doors. These two extra cleats support the 6×6 grade beam installed across the threshold of the doorway. Measure 8½ in. down from the top of the first cleat nailed to the inside of each corner post and make a mark. Measure from the bottom of the post to the mark and cut a 2×4 cleat to fit. Nail this second cleat to the first cleat with 3½-in.

Nail a 2×10 pressure-treated skirtboard to the corner posts at the rear of the foundation. Hold the board 1½ in. above the cleats.

Install a pressure-treated 6×6 grade beam between the two corner posts at the front of the foundation. It'll act as a threshold for the door.

(16d) galvanized nails. Then set all the posts back in their respective holes and temporarily screw a long 2×4 across them to hold them in place.

Cut a pressure-treated 2×10 to 11 ft. 5 in., which is the exact width of the rear wall, and nail it to the two corner posts at the rear of the foundation. Hold the top edge of the 2×10 skirtboard 1½ in. above the 2×4 cleats nailed to the post, making sure it is flush with the bottom wall plate.

At the front of the foundation, where the sliding door will eventually be hung, install a pressure-treated 6×6 grade beam to serve as a threshold between the two corner posts. Lay the timber on a bed of compacted processed stone. Make sure each end of the beam rests on the 2×4 cleats, then fasten the 6×6 beam to the posts with 3-in. decking screws.

Frame the walls

The barn's walls are framed with 2×4s and sheathed with grooved plywood siding. This is done while the walls lie flat on the ground; then they're lifted up and set between the posts. Begin by building the four sections that make up the two long sidewalls. For each section, cut a top and a bottom wall plate to 9 ft. 4¾ in. long. Next, cut six 2×4 studs to 74½ in. long. Use 3-in. decking screws to fasten the wall plates to the two end studs to create the rectangular wall frame. Check the frame for square by measuring the diagonals. Then install the remaining four studs and frame out the rough opening for the 2-ft. by 3-ft. aluminum window, following the drawing below. Make the opening 25¼ in. high by 36¼ in. long. Before moving on to the next

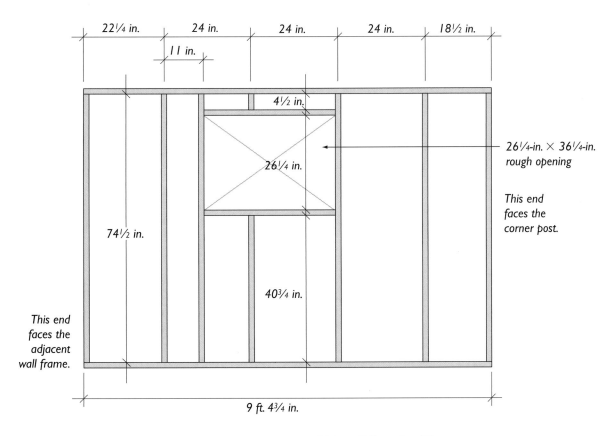

Sidewall Framing Plan
Each long side of the barn is made from two 9-ft.-4¾-in.-long wall frames, which are built entirely out of 2×4s. Note that each frame has a rough opening for a 2-ft. by 3-ft. window.

TRADE SECRET

It's nearly impossible to determine ahead of time exactly how tall to make each 4×4 post, because neither the ground nor the tops of the footings are perfectly level. For now, just take the uncut 12-ft.-long posts and set them in the footing holes. Come back after the walls are installed and trim the posts flush with the tops of the walls. I used a handsaw to cut through the 4×4s, but a reciprocating saw would have been faster.

WHAT CAN GO WRONG

If you live in a region that receives a lot of rain or if the building site doesn't drain water very well, be sure to use pressure-treated lumber for the bottom plates of the walls. Untreated lumber will quickly rot if it's constantly getting wet.

Build a 25¼-in.-high by 36¼-in.-long rough opening in the wall for a 2-ft. by 3-ft. sliding aluminum window.

Nail grooved plywood siding to the wall frame with 1½-in. siding nails. Cut the siding flush with the edges of the rough opening.

step, check the wall for square once again by measuring the diagonals.

When building the walls, lay out the studs for each one by starting from the same end. That way, the studs in one wall will align perfectly with the studs in the opposite wall. This is important because the studs will later be used to position the roof trusses.

Now prepare to cover the wall frame with ⅝-in.-thick grooved plywood siding. First, cross-cut a plywood sheet to 82 in. long, which is 4½ in. longer than the wall height. Later, when the wall is set between the posts, the overhanging

plywood will partially conceal the 2×10 skirtboard nailed across the inside of the posts. Apply a continuous bead of construction adhesive to the 2×4 wall frame. Lay the plywood on the frame, making sure its top end is flush with the top wall plate and its vertical edge overlaps the end stud by 3½ in. That extra bit of siding will cover the 4×4 corner post.

Nail the plywood to the wall frame with 1½-in. ring-shank siding nails spaced 10 in. apart. Finish installing the siding, making sure you cut it flush with the edges of the rough window opening. The last piece of siding extends past the end stud, too, but by only 1¾ in., which is half the width of the intermediate post. Now build three more identical sidewall sections. Make each one 77½ in. tall by 117¾ in. long, remembering to frame in the window in each section.

Next, build the two gable-end walls, making each one 77½ in. tall by 137 in. long. Note that the rear wall has a 25¼-in. by 36¼-in. rough window opening and the front wall has a 66-in. by 73-in. rough door opening. For each wall, cut a top and a bottom wall plate to 11 ft. 5 in. long. Then cut six 2×4 studs to 74½ in. long. Fasten the wall plates to the studs with 3-in. decking screws. Space the studs 24 in. on center. Cover the two gable-end walls with plywood siding. Make sure the plywood extends past both ends of each gable-end wall by 3½ in. to effectively cover the 4×4 corner posts. You're now ready to install the walls, starting at the rear of the barn.

Raise the walls

With the help of a friend, carry the rear wall to the back of the foundation. Slide it between the two rear posts, making sure its bottom wall plate is on top of the 2×4 cleats nailed to the posts. Raise the wall to a full upright position; if the fit is too tight, push out on the posts. As your helper holds the wall in place, move to the inside and fasten it

Slide the fully assembled rear wall between the two 4×4 corner posts and set it on top of the 2×4 cleats nailed to the posts.

Push the wall against the 4×4 posts from the outside and secure it by driving 3-in. screws through the end studs and into the posts.

TRADE SECRET

After backfilling around the 4×4 posts, use a garden rake to pull 2 in. of processed stone over the tops of the fiber-form tubes. That'll hide the tubes and ensure that they remain hidden even after the stone settles and compacts over time. Leave 4 in. to 6 in. of air space between the siding and the stone.

TRADE SECRET

Building a barn this size requires quite a bit of space—and not just for the barn. You'll also need plenty of space for cutting sheets of plywood, storing 12-ft.-long posts, and assembling 7-ft. by 12-ft. trusses. Set up a staging area right beside the building site so you won't have to carry the materials very far.

After the rear wall has been fastened, install one of the sidewall panels. Set it between the rear corner post and the intermediate post.

Fasten the sidewall to the corner post with 3-in. screws spaced 12 in. apart. Make sure the end stud is flush with the inside of the post.

to the posts with 3-in. decking screws spaced about 12 in. apart.

Next, install the first of the four sidewalls. These sections are much lighter and easier to handle than the rear wall. Slide the first one between the rear corner post and the intermediate post. Again, make sure its bottom plate sits directly on top of the 2×4 cleats. As your helper pushes the wall tightly against the posts from the outside, secure the wall by screwing through the end

studs. The wall is in the proper position when the end studs completely cover the sides of the post.

Install the second sidewall next to the one you just screwed in place. Check to make sure the plywood seam between the two wall sections fits together tightly at the intermediate post. Don't worry if the opposite end of the wall doesn't fully cover the corner post; that joint will be concealed later by corner boards. It's more important to create a tight seam at the intermediate post. After screwing the second sidewall in place, move to the opposite side of the foundation and install the two remaining sidewall sections.

Erect the front wall

The front wall is more than 11 ft. long, but it's very easy to lift into place because of the large doorway opening. However, before you can install it, you must attach double 2×4 sill plates to the

6×6 grade beam on each side of the doorway opening. These boards are required because the beam is 3 in. lower than the cleats that support the sidewalls. The sill plates will raise the front wall even with the sidewalls.

Cut four 34-in.-long sill plates from pressure-treated 2×4s. Stack two sill plates on top of the 6×6; make sure they're butted tightly against the corner post and centered on the 6×6. Fasten each plate with 3-in. decking screws. Fasten the other double sill plate to the opposite end of the grade beam.

Set the front wall between the corner posts and tilt it up into place. Screw through the end studs and into the posts, just as you did for the other walls, but also screw down through the wall's bottom plate and into the double sill plate. Move to the outside of the barn and fasten the plywood siding to each of the six posts with 1½-in. ring-shank siding nails spaced 10 in. to 12 in. apart.

Complete the wall framing

Finish up the wall framing by nailing a 2×10 skirtboard to the inside surface of the two sidewalls. Hold the 2×10 even with the top of the wall's bottom plate and drive 3½-in. (16d) galvanized nails into the edge of the bottom plate and into the 4×4 posts. The skirtboards add structural integrity to the walls and create a perimeter frame, which holds the processed stone in place.

The next step is to encase the six 4×4 posts in concrete to permanently hold them in position. Don't forget that at this stage the posts are just sitting on top of concrete footings; they're not secured within the fiber-form tubes. However, before you can encase them, you must make sure the walls are straight; otherwise, you'll never get the roof trusses to fit.

To straighten the walls, start by temporarily screwing a short 2×4 block to each end of the sidewall near the bottom of the wall. Stretch a

Fasten a double 2×4 sill plate to each end of the 6×6 grade beam. Cut the 34-in.-long sill plates from pressure-treated lumber.

Install the front wall last. Stand it on top of the double 2×4 sill plates, then tilt it between the two front corner posts.

Drive 3-in. screws through the bottom wall plate and into the sill. Also, screw through the end stud and into the corner post.

PRO TIP

If you rub wax or bar soap onto screw threads, they'll spin into the wood with much less strain on you and the drill's motor.

WHAT CAN GO WRONG

Assembling the trusses on the ground significantly reduces the amount of time it takes to frame the roof. However, you have to make sure you build all the trusses to exactly the same size. If one truss is even slightly taller than the others, it'll affect the installation of many other components, including the roof sheathing and the subfascia.

TRADE SECRET

The easiest, most accurate way to make square cross-cuts and angled miter cuts in framing lumber is with a power miter saw. However, most miter saws typically cut to only about 50 degrees, which means that you'll have to use a portable circular saw to make the 54-degree cuts needed for some of the rafter-truss components. One simple way to mark this miter cut is to lay a sliding-bevel square on top of a protractor and set the square's blade at a 54-degree angle. Then use the square to mark the angled line on the 2×4 rafters.

Nail a pressure-treated 2×10 skirtboard to the inside of the sidewalls. Hold the board flush with the top of the wall's bottom plate.

Use a taut string to check whether the wall is straight. Put a 2×4 block under each end of the string, then use another 2×4 as a gauge block.

Shovel half a bag of concrete into each footing hole. Be sure to distribute the concrete evenly around all four sides of the posts.

very taut string across the blocks and tie it off. Using a third 2×4 block as a measuring gauge, slip the block behind the string at several places along the wall. If there's not enough space for the block to slide behind the string, move the wall in toward the center of the barn. If the space between the string and the siding is greater than the thickness of the 2×4 block, push the wall out. The simplest way to move a wall is to slip a long 2×4 into the footing hole. Then simply pull back on the board to nudge the post.

After you've confirmed that all the walls are straight, mix up a bag of concrete in a wheelbarrow. Use a sharp utility knife to cut the fiber-form tube flush with the ground, then shovel one-half of the concrete into the tube, making sure you distribute it evenly on all sides of the post. Repeat the process for the other five posts, using half a bag of concrete per hole. Allow the concrete to cure overnight, then fill each hole to the very top with processed stone or gravel.

Roof Framing

It's never easy to frame a roof as big as this one, but using prebuilt trusses makes the job much simpler. Once you build the two gable-end trusses, it should take about two hours or so to frame the roof and another couple of hours to nail down the plywood roof sheathing.

You'll need at least four people to frame this roof: two on the ground handing up the trusses and two on top of the sidewalls setting the trusses in place. The ground workers also screw the trusses to the top wall plates from below. Note that it's important to place one truss directly over each wall stud. That way, the wall frames, posts, and footings will properly support the weight of the roof.

Roof-truss assembly

A typical gable roof truss has only three major parts: two angled rafters and a horizontal bottom chord. Each of the nine standard trusses in this gambrel barn is framed with seven pieces: four rafters, two vertical braces, and one bottom chord. The top two rafters meet at the peak and form a relatively shallow 6-in-12 slope. The roofline then breaks down sharply where the second pair of rafters forms a very steep 24-in-12 slope.

The rafters and braces are cut from 2×4s, but for added strength the bottom chord is a 2×6. The boards that make up the trusses are held together with plywood gusset plates glued and nailed to each side of the truss. In the previous two chapters, the trusses were assembled on the plywood floor deck. Unfortunately, you can't use that technique here because the barn doesn't have a plywood floor. Instead, you must assemble the

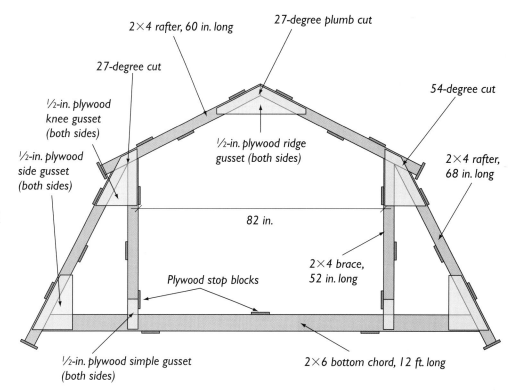

Roof-Truss Assembly
Gambrel roof trusses are slightly more difficult to build than standard gable trusses. Each one has four 2×4 rafters, two 2×4 vertical braces, and one 2×6 bottom chord. One truss is assembled on the ground and used as a template for the remaining trusses. A series of plywood stop blocks are screwed to the template truss and used to hold the parts in proper alignment. Plywood gusset plates are glued and nailed across the joints to hold each truss together.

trusses on the ground, in the driveway, or on another large, flat surface.

For each truss, cut two 2×4 top rafters to 60 in. long. Trim one end of each rafter at a 27-degree angle and cut the other end square.

✔ **According to Code**

There are only six 4×4 posts supporting this 12-ft. by 20-ft. storage barn. The building inspector initially balked at this detail, but he later approved it when he saw that a 2×10 skirtboard reinforced the walls. Be aware that the local code in your area may require you to beef up the structure by using 6×6 posts.

TRADE SECRET

When it comes time to install the roof trusses, employ a tag-team approach: Have one person stay up on the wall and set a truss directly over a wall stud. Then, as you drive two screws from below, have him or her stand on top of the bottom chord to prevent the truss from sliding out of position. Note that it's necessary to drive the screws at a slight angle in order to miss the stud.

Make a template truss by screwing ½-in. plywood stop blocks to the ends and edges of the rafters and bottom chord.

Then cut two 2×4 lower rafters to 68 in.; miter-cut one end at a 54-degree angle. Next, cut the two vertical 2×4 braces to 52 in. Again, trim one end at a 27-degree angle and cut the other end square. Finally, cut the 2×6 bottom chord to 12 ft. long, mitering both ends at a 27-degree angle.

For each truss, you must also cut 14 gusset plates from ½-in. ACX plywood. Cut the two triangular ridge gussets to 9½ in. high by 24 in. wide, and cut eight side gussets to 13 in. by 18 in. In addition, you'll need four 3½-in.-wide by 12-in.-long simple gussets to secure the vertical braces to the bottom chord.

Next, cut twenty-five 3-in. by 6-in. stop blocks from ½-in. plywood. Assemble the parts to make one truss, but attach gusset plates to only one side, using construction adhesive and 1-in. roofing nails. Use 10 nails per gusset. Temporarily fasten the stop blocks to the edges of the rafters, braces, and bottom chord with 1¼-in. screws. Refer to the drawing on p. 183 for the exact placement of the stop blocks. You can now use this truss as a template for assembling the others.

Lay the parts for the next truss directly on top of the template truss. The stop blocks will hold each piece in place, but check to make sure the

Set the truss parts on top of the template truss. The plywood stop blocks will hold all the boards in proper alignment.

joints fit together tightly. Glue and nail on the gusset plates with construction adhesive and 1-in. roofing nails. Flip over the truss and attach gussets to the other side. After assembling the last truss, add gussets to the template truss and unscrew the stop blocks.

Build the gable-end trusses

In addition to the nine standard trusses, you'll need to build two gable-end trusses. Start by assembling two trusses to the same dimensions as the standard trusses. Next, add five more vertical 2×4 braces, spacing them equally between the two original braces. These extra pieces provide solid support for the grooved plywood siding. Also, screw a 10-ft.-long 2×4 shoe plate to the bottom chord. When the truss is installed, the shoe plate will rest on top of the gable-end wall.

Glue and nail plywood gusset plates to only one side of the truss, then flip over the truss and sheathe it with plywood siding. Make sure the

Use construction adhesive and 1-in. roofing nails to fasten the ½-in. plywood gusset plates to each side of the truss.

The gable-end trusses are similar to the standard roof trusses, except that they're covered with grooved plywood siding.

Allow the plywood siding to extend past the top of the truss, then use a circular saw to trim the siding flush with the roof rafters.

Attach rake boards to the gable-end truss with 1½-in. siding nails. Cut each 5-in.-wide rake from a roughsawn cedar 1×6.

seams between the plywood pieces fall on the center of a 2×4 brace. Secure the plywood with construction adhesive and 1½-in. ring-shank siding nails. Run the siding long, then use a circular saw to trim the plywood flush with the rafters.

Next, make the rake boards by ripping rough-sawn cedar 1×6s to 5 in. wide. Miter the rakes to match the roof slope and cut them to extend about 6 in. past the rafter tail. That extra bit of rake will be trimmed off after the fascia is installed. Position them to project ½ in. above the top edge of the rafters; that raised lip will hide the edge of the plywood roof sheathing. Secure the rake boards to the truss with 1½-in. siding nails. Finally, nail 1×2 cedar boards—called rake trim—to the rake boards.

Carry one of the gable-end trusses inside the barn and set it upside-down on top of the side-walls. Make sure its plywood surface faces toward the center of the barn. Grab the peak of the truss and rotate it upward. Have two helpers on ladders

PRO TIP

To nail into the center of each rafter, mark layout lines every 24 in. on center on each sheet of plywood, then drive the nails on the line.

TRADE SECRET

It's important to install the perforated aluminum soffit vent before covering the lower roof slope with plywood. Otherwise, you won't be able to nail the vent in place. Install the vent and secure its upper edge with 1-in. roofing nails. Drive the nails through the vent's top flange and into the plywood siding; space the nails 12 in. apart.

TRADE SECRET

You'll need the help of at least two friends to sheathe the roof of this barn. But if your friends can help for only a short while, use the time wisely. Secure each sheet with only four or five nails, then quickly move on to the next sheet. After your friends leave, you can finish nailing off the plywood sheets alone.

Set the gable-end truss on top of the end wall. Slowly push the truss to an upright position; be careful that it doesn't tip over.

Fasten the gable-end truss by driving 3-in.-long decking screws down through the 2×4 shoe plate and into the top wall plate.

+ SAFETY FIRST

When framing a roof, it's a good idea to nail a temporary 2×4 brace across the trusses to hold them upright and prevent them from accidentally falling over. Set the brace horizontally across the trusses, positioning it near the upper end of the lower rafters. Nail the end of the brace to the gable-end truss. As you install each truss, secure it by nailing through the brace and into the rafter.

slowly push the truss to a full, upright position. Align the 2×4 shoe plate on the truss with the top wall plate. Secure the truss by driving 3-in. decking screws down through the shoe and into the top wall plate; drive one screw every 12 in. Install the second gable-end truss in a similar manner.

Install the roof trusses

To install the nine standard roof trusses, you must lift each one onto the walls from the outside. Fortunately, the trusses aren't all that heavy, but at 7 ft. tall by 12 ft. long, they're a little unwieldy. Enlist the help of at least four people: Place two workers inside on ladders and two outside to handle the trusses.

Lift up the first truss, set it on top of the sidewall, and slide it across to the helper at the opposite wall. Set the truss directly over the first wall stud. Line up the very end of the bottom chord with the outside edge of the wall's top plate. Drive two 3-in. screws up through the top plate and

Lift the prefabricated roof trusses onto the sidewalls from the outside. Altogether there are nine trusses used to frame the roof.

Set each truss directly over a stud, then drive two 3-in.-long screws up through the top plate and into the bottom chord of the truss.

into the bottom of the truss. Don't fasten the other end of the truss just yet. Install the remaining trusses, making sure you place one over each stud.

Now begin fastening the trusses to the opposite sidewall, starting with the truss in the middle. Using the bottom chord as a guide, push the wall in or out to align the end of the chord with the top wall plate, then fasten the truss with two 3-in. screws driven from below. With the middle truss holding the wall straight, attach the remaining trusses.

Next, cut a long 2×4 subfascia to span the distance across the tails of the upper rafters. Mark the 2×4 with the location of the rafters, which are spaced 24 in. on center. Hold the 2×4 subfascia against the ends of the rafter tails, align each tail with the layout mark, and nail the subfascia to the ends of the tails with 3½-in. (16d) nails.

Fasten a long 2×4 subfascia to the ends of the rafters on the upper roof slope; drive two 3½-in. nails into each rafter tail.

IN DETAIL

When installing ½-in. plywood sheathing on the lower slope of the gambrel roof, it's important that you tuck the sheet underneath the overhanging edge of the upper slope. Lay the plywood across the rafters, then raise its lower edge and slide it tightly against the overhang. Securely nail the sheet to each rafter with 1½-in. (4d) nails.

IN DETAIL

There's a detail on this gambrel roof that you seldom see on outbuildings. It has a fascia board installed along the edge of the upper roof slope. Ordinarily, the roof shingles are simply bent to cover the joint between the upper and the lower slopes. That not only looks bad but also creates a fault line along the joint, where the shingles will be more likely to crack and wear out.

Sheathe the roof

You'll need 15 sheets of ½-in.-thick exterior-grade plywood (BCX or CDX grade) to cover this gambrel roof. Sheathe the upper roof slope first, then cover the lower slope.

Start with a full 4-ft. by 8-ft. sheet, laying it on top of the rafters and aligning its lower edge flush with the 2×4 subfascia. Butt the end of the sheet tightly against the rake board, then fasten the plywood to the rafters and subfascia with 1½-in. (4d) nails spaced 10 in. apart.

Install another full sheet of plywood, followed by a half sheet. At the peak, rip the plywood to fit, leaving it 1½ in. short of the peak on each side of the roof; that narrow space will be covered later by the ridge vent.

Cover the roof frame with ½-in.-thick exterior-grade plywood. Fasten the sheathing with 1½-in. nails spaced 10 in. apart.

Hold the soffit vent on the subfascia with a strip of aluminum drip-edge flashing. Secure the L-shaped flashing with 1-in. roofing nails.

When the upper roof slope is covered, start sheathing the lower slope, beginning with a full 4-ft. by 8-ft. sheet of plywood. Slide the sheet over the rafters and fit it tightly underneath the overhanging rafter tails of the upper roof slope. Continue covering the upper portion of the lower roof slope with plywood, but don't sheathe the lower section just yet. You must install the soffit vent first.

Nail a 2×4 subfascia to the ends of the rafter tails, then slip a 6-in.-wide section of perforated aluminum soffit vent underneath the roof overhang and press it tightly against the underside of the rafter tails. Use a length of aluminum drip-edge flashing to hold the bottom edge of the vent to the subfascia. Place the L-shaped flashing over the vent and fasten it to the subfascia with 1-in. roofing nails. The reason it's necessary to use flashing is that there's not enough room below the rafter tails to nail or screw the vent in place.

Fasten the top edge of the soffit vent to the plywood siding with 1-in. roofing nails spaced about 12 in. apart. When the vent is installed, you can finish nailing down the plywood roof sheathing,

Add the fascia

Cut 5-in.-wide fascia boards from roughsawn cedar 1×6s and nail them to the subfascia with 2½-in.-long siding nails. Make sure the upper edge of the fascia is even with the plywood sheathing. Note that you'll need to cut two fascia boards to span the distance. Join the two boards together with a simple butt joint. Drill ⅛-in.-dia. pilot holes to prevent the nails from splitting the ends of the boards. Use a handsaw to trim the rake board on the gable-end truss flush with the fascia. Nail a 1×2 cedar fascia trim to the fascia. Again, keep the top edges flush. Then trim the overhanging end of the 1×2 rake trim even with the fascia trim.

Nail the fascia board to the subfascia with 2½-in.-long siding nails. Cut the 5-in.-wide fascia from a roughsawn cedar 1×6.

Use a handsaw to trim the wide cedar rake board on the gable-end truss until it is flush with the fascia board.

Nail a 1×2 cedar fascia trim to the fascia board. Then trim the overhanging end of the 1×2 rake trim so it is even with the fascia trim.

PRO TIP

The best time to begin shingling the roof is in the early morning or late afternoon, when the weather is a little cooler.

TRADE SECRET

Many roofers snap chalklines while shingling to help keep shingle courses straight. These architectural-style shingles have a built-in alignment guide, and the bottom edge of each shingle is aligned flush with the horizontal edge of the upper lamination. Of course, this is all referenced from the first course; if that's off, then the whole roof will be off. I didn't snap chalklines, but I did measure up from the eave every few courses just to make sure I wasn't straying off course.

WHAT CAN GO WRONG

The last course of roof shingles on the lower slope is tucked underneath the edge of the upper slope. Because it's the last course, the nail heads are exposed to the weather. Over time, they'll rust and stain the roof. Prevent this by covering each nail head with roofing cement, then sprinkle the cement with loose granules taken from a shingle bundle. This trick will help the cement blend in with the shingles.

Fasten each starter-course shingle with three 1¼-in. roofing nails. Butt the shingles together tightly, but don't overlap them.

Roofing

You'll need four and half squares (450 sq. ft.)—about 18 bundles—of architectural-style asphalt shingles to cover the roof of this Gambrel Storage Barn. You'll also need 130 linear ft. of three-tab shingles (four bundles) for the starter course and 20 linear ft. of hip-and-ridge shingles (one bundle) to cover the ridge vent.

Install the starter course

The starter course consists of three-tab shingles nailed along the ends and edges of the roof. This course protects the perimeter of the roof and ensures that the plywood sheathing doesn't show between the architectural shingles. To completely cover the plywood, you must install the starter-course shingles upside-down, with the tabs facing toward the center of the roof.

Start by snapping a chalkline 11½ in. from the bottom edge and each end of the roof. Because the shingles are 12 in. wide, this will produce a ½-in. drip-edge overhang. Set a three-tab shingle on one of the vertical chalklines that runs up to the peak, fastening the shingle with three 1¼-in. roofing nails driven through the asphalt sealing strip. Install another shingle vertically on the opposite end of the roof.

Next, install starter-course shingles along the bottom edge of the roof, setting them on the chalkline and securing each with three nails. When you reach the end of the roof, cut the last shingle so it fits tightly against the vertical shingle without overlapping it.

Shingle the roof

Start at one end of the roof and set the first architectural-style shingle directly on top of the starter course. Make sure its end and lower edge are flush with the starter shingle below, then fasten the shingle with four 1¼-in. roofing nails. Drive one nail 1 in. from each end of the shingle and space the other two nails 12 in. in between. Continue laying full-size shingles until you reach the end of the roof, being careful to overlap all the vertical seams in the starter course by at least 6 in. Then cut the last shingle to length, allowing for a ½-in. overhang at the end.

Start the second course by first cutting 6 in. off a shingle; that will automatically stagger the seams between the first and second course by 6 in. Place the first shingle in the second course flush with the end of the roof and align its lower edge to produce a 5-in. exposure to the weather, then secure it with four 1¼-in. roofing nails. The easiest way to maintain the correct exposure is to align the bottom edge of the shingles with the horizontal lines formed by the upper laminations on the shingles below. Continue nailing down full-length shingles in this manner. As you make your way up the roof, add starter shingles, as needed, along the vertical ends of the roof.

To start the third course, cut 12 in. off a shingle; this 24-in.-long shingle will maintain the 6-in. offset between the seams. Repeat this pattern for each subsequent course. Start the fourth course with an 18-in. shingle, the fifth course with a 12-in. shingle, and the sixth course with a 6-in. shingle. Start the seventh course with a full-size shingle and begin the pattern all over again. When you get to the last (uppermost) course, tuck the shingles underneath the upper-slope overhang, nailing them as close as possible to the fascia.

Use the same roofing technique to shingle the upper slopes. At the peak, trim the shingles flush with the edge of the plywood. Then install a flexible polyester-fiber ridge vent. Unroll the vent along the ridge and fasten it to the roof with 2-in.-long roofing nails. Don't pound the nails all the way in or you'll crush the vent. Cover the vent with hip-and-ridge shingles, making sure you maintain a 5-in. exposure to the weather. Fasten each shingle with two 2-in. roofing nails.

Cover the roof with architectural-style shingles. Maintain a 5-in. exposure to the weather and secure each shingle with four nails.

Set the windowsill in the rough opening. Note that the ends of the 5/4 by 6-in. cedar sill are notched to extend past the opening.

Windows and Exterior Trim

This barn has five 2-ft. by 3-ft. aluminum sliding windows that provide plenty of natural light and fresh air. There are two windows in each sidewall and one in the rear wall. The ready-to-install units have an integral mounting flange, insect screens, and built-in weep holes that drain water trapped in the lower track. However, before you can set the windows in place, you must install the windowsills.

Tilt the aluminum window into the opening, placing the top end first. Press it tightly against the siding and secure it with screws from the inside.

TRADE SECRET

This barn floor is made from processed stone, which is a mixture of gravel and stone dust. After raking the stone smooth, sprinkle the surface with water and then pound it flat with a gas-powered plate compactor. The wet dust will form a mortarlike slurry that'll bond the stones together, creat- ing a very flat, hard surface. Over time, the stones may settle in spots or get scraped off. When that happens, spread a fresh 2-in. to 4-in. layer of stone over the floor, dampen it, and compact it.

WHAT CAN GO WRONG

Whenever you build doors out of plywood, there's always a chance that they may warp. That's because the cross- grain plies absorb moisture and expand and contract at different rates. To deter warping, apply at least two coats of paint or stain to all surfaces of the plywood panel, especially the porous edges. Reinforce the front and rear of the door with a wood frame made from redwood or cedar. Also, keep the space beneath the door free of debris that can trap moisture.

Use 1½-in. siding nails to attach a vertical casing to each side of the window. Cut the side cas- ings from a roughsawn cedar 1×4.

Nail a horizontal trim piece, called an apron, to the wall directly below the windowsill. Cut the apron from a cedar 1×4.

Install the windows

For each window, start by cutting a 44¼-in.-long windowsill from a 5/4 by 6-in. cedar board. Then cut a 4⅛-in.-wide by 4-in.-deep notch in each end. Set the sill in the window opening and, if necessary, shim it level. Attach it to the 2×4 rough sill with six 2½-in.-long siding nails.

Next, set each aluminum window in its open- ing and press it tightly against the siding. As you hold it in place, have a helper inside the barn slide the window tightly against the wall stud on one side of the opening and secure it by driving two 2-in. screws through the mounting holes in the window's side jambs and into the wall stud. Slip a thin wood shim between the opposite side of the window and the wall stud and secure it with 2-in. screws.

Trim out the windows

Cut two 1×4 cedar side casings to 24¾ in. long. Set each one on top of the sill, making sure it overlaps the aluminum window frame by ½ in. Attach the casings to the sidewall with 1½-in. siding nails. Cut a 1×4 cedar head casing to 43¼ in. long and set it on top of the side casings.

Make sure it overhangs each casing an equal amount, then nail it in place.

Next, make the apron, which is the horizontal trim piece running underneath the sill. Cut a 1×4 cedar board to 43¼ in. long and nail it to the sidewall below the windowsill. Make sure the ends of the apron align with the outside edge of the side casings.

Add the stone floor

Before installing the doors, order a truckload of processed stone from a masonry supplier. You'll need between 4 cu. yds. and 6 cu. yds. Have the load dumped right at the doorway opening, then use a wheelbarrow to transport the stone into the barn. Put down a 4-in. to 6-in. layer of stone and rake it smooth.

Door Installation

The doors (which are built in the sidebar on pp. 194–195) are hung with traditional sliding-door hardware. Two roller assemblies are mounted on the top of each door. Each assembly has two round rollers. A round tubular steel track is bolted over the doorway opening, then the rollers are slipped into the track. This type of hardware is called a cannonball hanger. Trolley hangers, which have wheels similar to those on roller skates, are also popular.

Attach the sliding-door hardware

Start by cutting two ⅞-in.-wide by 1½-in.-deep slots in the top edge of each door. Position the notches 5½ in. from the edges of the doors. Cut the notches using a handsaw, a hammer, and a sharp chisel.

Hold the rear mounting plate against the back of the door and bore two ⅜-in.-dia. bolt holes through the doorframe. Pass the two carriage bolts through the holes. Insert the square nut from one

Lay the front mounting plate on the door, trapping the square nut in the notch. Tighten the two carriage bolts with a socket wrench.

Slide the metal mounting clips into the keyhole-shaped slots cut in the rear of the tubular-steel door track.

of the roller assemblies into the notch cut in the door. Place the front mounting plate on the door, trapping the square nut in the notch. Thread hex nuts onto the bolts and tighten them with a socket wrench.

Next, use a hacksaw to cut 12 in. off the 12-ft.-long tubular steel track. Note that the track has a

BUILDING SLIDING DOORS

The sliding doors on this shed resemble board-and-batten doors, but they are actually made from grooved plywood siding. Each 34-in. by 77-in. plywood panel is sandwiched between a front and a rear perimeter frame, then three 1×2 vertical battens are applied to the front surface. Cross battens, cut from 1×6 cedar, are screwed to the backs of the doors to add strength and deter warping. Here are the basic steps for building each door:

1. Cut the plywood door panel to size, then cut the four parts of the perimeter face frame from roughsawn cedar 1×4s. Cut two 35½-in.-long horizontal rails and two 70-in.-long vertical stiles.

2. Fasten the face frame to the plywood panel with construction adhesive and 1¼-in. (3d) galvanized nails. Note that the rails overhang the edge of the panel by 1½ in. to create an over-

Apply a bead of construction adhesive to the backs of the 1×4 cedar face-frame parts. Nail the parts to the plywood door panel.

Create the look of a board-and-batten door by attaching 1×2 battens to the door panel. Place the battens over the plywood grooves.

Glue and screw a perimeter frame to the rear of the plywood door panel. Cut each 4½-in.-wide frame part from a cedar 1×6.

lapping wind-stop joint where the two doors meet. Create a similar overlap on the edge of the mating door.

3. Use adhesive and nails to attach the three vertical 1×2 cedar battens to the door face. Center each batten directly over one of the grooves milled in the plywood.

4. Flip over the door. Rip five cedar 1×6s to 4½ in. wide. From these boards, cut two 34-in.-long rails, two 68-in.-long stiles, and one 25-in.-long center rail. Fasten the parts to the perimeter of the door panel with panel adhesive and 1¾-in. decking screws spaced no more than 10 in. apart.

5. Cut diagonal cross battens to fit between the horizontal rails screwed to the rear of the door. Attach the battens with adhesive and 1¾-in. decking screws.

Reinforce the perimeter frame with diagonal cross-buck battens. Attach the battens with adhesive and 1¾-in. decking screws.

Install the tubular door track over the doorway opening. Make sure it's level, then fasten it with 2-in.-long lag screws.

series of keyhole-shaped slots cut in it. Insert a metal mounting clip in each of the six slots, using a hammer to gently tap the clip all the way in.

Raise the track over the doorway and fasten one end by driving a ⅜-in.-dia. by 2-in.-long lag screw through the hole in the first mounting clip. The center of the rail should be 76½ in. above the top of the 6×6 grade beam. Make sure the screw passes through the plywood siding and into the 2×4 wall frame. If there's no framing member directly behind the door track, add blocking. Use a 4-ft. level to make sure the track is perfectly level, then drive a lag screw through the mounting clip at the opposite end of the track. Check the track for level one more time, then screw the remaining four mounting clips to the wall.

Hang the doors

These doors are very easy to hang, but it takes two people to get them up onto the track. Start by setting up a ladder near the end of the tubular steel track. With the help of a friend, bring over one of

PRO TIP

To keep the sliding doors rolling smoothly, apply a little multipurpose lithium grease to the roller assemblies once a year.

TRADE SECRET

Over time, the guide rollers at the bottom of the doorway can wear through the paint or stain finish applied to the bottom door rails. In some cases, the rollers can even carve a groove into the soft cedar rails. To prevent this damage, protect the door with a 2-in.-to-3-in.-wide strip of flat-stock aluminum or stainless steel. Cut each strip as wide as the doors. Drill and countersink screw holes in the strips, spacing them about 12 in. apart. Then, align the center of the strip with the center of the guide roller and attach it with 1¼-in.-long aluminum or stainless steel flathead screws.

IN DETAIL

The 2×4s used for the subfascia are nearly 20 ft. long. If you can't buy or handle boards that long, span the distance with two 2×4s. Cut the boards so the splice falls in the center between two rafter tails. Then cut a 2×4 block to fit between the tails. Set the block behind the joint and fasten it to the subfascia with construction adhesive and twelve 3-in. decking screws.

Lift up the door and guide the two roller assemblies into the tubular track. Repeat this procedure to hang the second door.

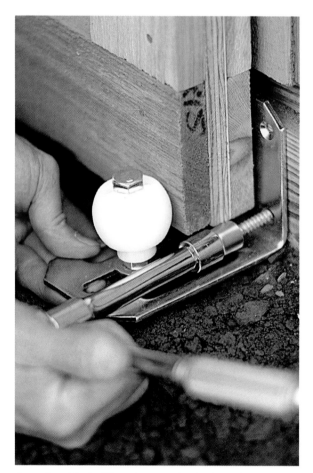

Attach guide-ball brackets to the 6×6 grade beam with ⅜-in.-dia. by 2-in. lag screws. The brackets hold in the door bottoms.

the doors and lean it against the front wall of the barn. Climb up the ladder. As your helper raises the door from below, guide the roller assemblies into the track. Slide the door all the way to the other end of the track and repeat the process to hang the second door.

Check to make sure the doors slide smoothly and don't rub against the barn. Also, bring the doors together to see whether they meet squarely. You may have to raise or lower one of the elevation-adjustment nuts on the roller assemblies to get the doors to hang straight and meet flush. If a door rubs against the siding, loosen the lateral-adjustment nuts on the roller assemblies. Use a hammer to tap the interior mounting plate until the door moves away from the barn. Tighten the nuts and test the door.

Once you're satisfied with the way the doors open and close, install the two lower guide-ball brackets. These pieces prevent the bottoms of the doors from swinging away from the barn. One bracket is needed for each door.

Open the doors all the way until they're flush with the edge of the doorway opening. Mount a

Once the doors are up, nail cedar corner boards to the barn's four wall corners. Fasten the boards with 2½-in. siding nails.

guide-ball bracket 6 in. from the outer edge of each door, fastening the brackets to the 6×6 grade beam with ⅜-in.-dia. by 2-in. lag screws. Loosen the nut on the underside of the bracket and move the guide ball to within ⅛ in. of the door, then tighten the nut.

Finally, install 1×4 roughsawn cedar corner boards on the four corners of the barn. Miter-cut the top ends of the boards to match the slope of the roof and fasten them with 2½-in.-long siding nails spaced 12 in. apart. Note that the corner boards can't be installed earlier because they will interfere with sliding the roller assemblies into the tubular steel track.

Bonus Storage

One benefit of building this storage barn is that its double-sloping gambrel roof offers plenty of overhead storage space. Above the ceiling joists is an "attic" area that's about 6 ft. high by 6½ ft. wide by 18 ft. long. It's an ideal out-of-the-way place for storing seldom-used items, such as old tools, camping gear, holiday decorations, suitcases, and boxes of seasonal clothing.

After installing a plywood floor, the challenge is finding the best way to access the space. A fold-down attic staircase is a safer and more secure alternative than a simple ladder. The cleverly designed folding staircase is mounted on a pull-down door panel equipped with spring-loaded brackets. The hinged stairs fold flat against the panel and the brackets extend when the unit is lowered. To access the storage space, you simply pull down the door panel and unfold the stairs.

A fold-down attic staircase provides an easy and safe way to gain access to the overhead storage space above the ceiling joists.

Resources

Plans for Sheds in this Book

To order a set of building plans for the Lean-To Shed Locker, send a check for $19.95 to:
RWS Company
105 South St.
Roxbury, CT 06783
(860) 350-1089

To order a set of building plans for the Saltbox Potting Shed, Colonial-Style Shed, or Gambrel Storage Barn, send a check for $29.95 per plan to:
Better Barns
126 Main St. South
Bethlehem, CT 06751
(203) 266-7989
www.betterbarns.com

Mail-Order Shed Plans
Barn Plans
213 2nd Texas Rd.
St. George, SC 29477
(843) 563-7267
www.barnplan.com

Donald J. Berg, AIA
Box 698
Rockville Centre, NY 11571
(516) 766-5585
www.abetterplan.com

Georgia-Pacific
133 Peachtree St., N.E.
Atlanta, GA 30303
(404) 652-4000
www.gp.com

Hometime
4275 Norex Dr.
Chaska, MN 55318
(888) 972-8453
www.hometime.com

To order a set of building plans for the post-and-beam shed shown on p. 13, send a check for $50 to:
Mercurial Editorial
Box 94
Maryknoll, NY 10545

Prebuilt Shed Kits
Backyard City
5910 Mt. Moriah #113-240
Memphis, TN 38115
(877) 233-9829
www.backyardcity.com

Handy Home Products
6400 E. 11 Mile Rd.
Warren, MI 48091
(800) 221-1849
www.handyhome.com

Summerwood Inc.
733 Progress Ave.
Toronto, Ontario M1H 2W7
(800) 663-5042
www.summerwood.com

Walpole Woodworkers
767 East St.
Walpole, MA 02308
(800) 343-6948
www.walpolewoodworkers.com

Pressure-Treated Lumber

Arch Wood Protection
1955 Lake Park Dr., Suite 250
Smyrna, GA 30080
(770) 801-6600
www.wolmanizedwood.com

Chemical Specialties
200 E. Woodlawn Rd., Suite 250
Charlotte, NC 28217
(800) 421-8661
www.treatedwood.com

Georgia-Pacific
133 Peachtree St., N.E.
Atlanta, GA 30303
(404) 652-4000
www.gp.com

Osmose Wood Preserving
980 Ellicott St.
Buffalo, NY 14209
(716) 882-5905
www.osmose.com

Southern Pine Council
Box 641700
Kenner, LA 70064
(504) 443-4464
www.southernpine.com

Cedar

Cedar Shake & Shingle Bureau
Box 1178
Sumas, WA 98295
(604) 820-7700
www.cedarbureau.org

Western Red Cedar
   Lumber Association
1200–555 Burrard St.
Vancouver, BC V7X 1S7
(604) 684-0266
www.wrcla.org

Redwood

California Redwood Association
405 Enfrente Dr., Suite 200
Novato, CA 94949
(888) 225-7339
www.calredwood.org

Concrete, Mortar, and Masonry Supplies

Bonsal
P.O. Box 241148
Charlotte, NC 28224
(704) 525-1621
www.bonsal.com

Sakrete
1402 N. River St.
Portland, OR 97227
(800) 245-3833
www.sakreteconcrete.com

Precast Dek-Block Pier Blocks
DekBrands
P.O. Box 14804
Minneapolis, MN 55414
(952) 746-7562
www.deckplans.com

Shingle Shield Protector Strips
Chicago Metallic
4849 Austin Ave.
Chicago, IL 60638
800-323-7164
www.chicago-metallic.com

Metal Framing Connectors
Simpson Strong-Tie
P.O. Box 1568
San Leandro, CA 94577
(800) 999-5099
www.strongtie.com

Ramp-Building Kit
Ramparts
Highland Group Industries
 31200 Solon Rd., Suite 1
Solon, OH 44139
(800) 234-6992
www.ramparts.com

Fiber-Cement Siding
CertainTeed
P.O. Box 860
Valley Forge, PA 19482
(610) 341-7000
www.certainteed.com

James Hardie Building Products
26300 La Alameda, Suite 250
Mission Viejo, CA 92691
(888) JHARDIE
www.jameshardie.com

Dura-Slate Roof Shingles
Royal Building Products
1 Royal Gate Blvd.
Woodbridge, Ontario L4L 8Z7
(905) 264-0701
www.royalbuildingproducts.com

Exterior Wood Finishes
Samuel Cabot Inc.
100 Hale St.
Newburyport, MA 01950
(800) US-STAIN
www.cabotstain.com

Aluminum Sliding Windows
PGT Industries
1070 Technology Dr.
Nokomis, FL 34275
(877) 550-6006
www.pgtindustries.com

Sliding Barn Door Hardware
CannonBall: HNP
P.O. Box 835
Beloit, WI 53512
(800) 766-2825
www.cnbhnp.com

Gambrel-Roof Framing Hardware
Lee Valley Tools
P.O. Box 1780
Ogdensburg, NY 13669
(800) 871-8158
www.leevalley.com
(Fast-Framer Kit, Item AC503)

Solar-Powered Photovoltaic Cells
Real Goods
360 Interlocken Blvd.
Suite 300
Broomfield, CO 80021
(800) 762-7325
www.realgoods.com

Solar Living Institute
P.O. Box 836
Hopland, CA 95449
(707) 744-2017
www.solarliving.org

Index